SMP **11-16**

Book G8
new edition

9902

CAMBRIDGE
UNIVERSITY PRESS

PUBLISHED BY THE PRESS SYNDICATE OF THE UNIVERSITY OF CAMBRIDGE
The Pitt Building, Trumpington Street, Cambridge, United Kingdom

CAMBRIDGE UNIVERSITY PRESS
The Edinburgh Building, Cambridge CB2 2RU, UK   www.cup.cam.ac.uk
40 West 20th Street, New York, NY 10011–4211, USA   www.cup.org
10 Stamford Road, Oakleigh, Melbourne 3166, Australia
Ruiz de Alarcón 13, 28014 Madrid, Spain

First published 1987
Fourth printing 1992
Second edition 1994
Reprinted 1995, 1999

Printed in the United Kingdom at the University Press, Cambridge

A catalogue record for this book is available from the British Library

ISBN 0 521 45747 5 paperback

Illustrations by Chris Evans and David Parkins
Diagrams by Marlborough Design
Photographs by Wales Tourist Board (pp. 60, 61), Graham Portlock (p. 72), Paul Scruton
(p. 74 top), The Architectural Press (p. 74 centre left), Shell (p. 74 centre right),
and Nigel Luckhurst.

Front cover photograph © NASA/Science Photo Library.

For permission to reproduce parts of leaflets thanks are due to the Bank of Scotland,
Barclays Bank, Midland Bank and National Westminster Bank.

Notice to teachers
It is illegal to reproduce any part of this work in material form (including photocopying
and electronic storage) except under the following circumstances:
(i) where you are abiding by a licence granted to your school or institution by the
Copyright Licensing Agency;
(ii) where no such licence exists, or where you wish to exceed the terms of a licence,
and you have gained the written permission of Cambridge University Press;
(iii) where you are allowed to reproduce without permission under the provisions of
Chapter 3 of the Copyright, Designs and Patents Act 1988.

Contents

Three in a line

You need a calculator and graph paper.

Draw a straight line on the graph paper.
Number it like this.

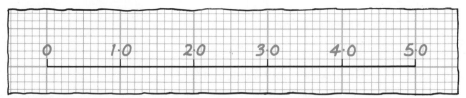

Players take it in turn to have a go.
Decide who starts.
When it is your turn, you pick two numbers from this list.

You divide one number by the other.
Then you mark the answer on the number line,
and write the division by it.

One player marks above the line,
the other player marks below.

If your answer is more than 5, you miss a go.

You are not allowed to choose a division that has already been used.

The winner is the first player to get 3 answers together,
without any of the other player's answers between.

For example, here is a win for the top player.

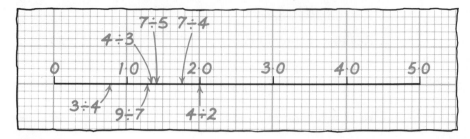

Try...

A different game.
Mark the graph paper from 0 to 1.
An answer over 1 won't count.

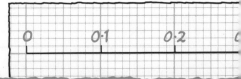

1 Volumes

A Cubes

> This is a box of cube sugar.
>
>

> These are the cubes of sugar taken out of the box.
>
>

A1 There are 3 layers of sugar cubes in a box.
  (a) How many cubes are there in 1 layer?
  (b) How many sugar cubes are there altogether in a box?

A2 (a) How many grams do the cubes weigh altogether?
  (b) About how much does 1 cube weigh?

> These cubes are 1 cm cubes. They are smaller than a sugar cube.

> These are the dimensions of the inside of the sugar box.
>
>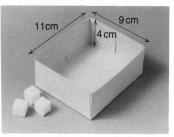

A3 (a) How many **centimetre** cubes could you get in 1 layer in the sugar box?
  (b) How many layers of centimetre cubes could you get in?
  (c) How many centimetre cubes could you fit in the sugar box altogether?

The number of centimetre cubes you can get in a box
is called its **volume.**

The volume of the sugar box is 396 centimetre cubes.

We usually say **cubic centimetres** instead of centimetre cubes.

You can write the volume in several different ways.

396 cubic cm   396 cu.cm.   396 cm^3 396 cc

A4 (a) How many centimetre
cubes could you get
in 1 layer in this box?

(b) How many layers
of cubes could
you get in?

(c) What is the volume
of the box?

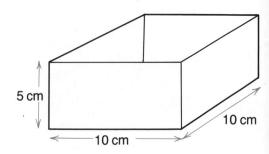

A5 What is the volume of each of these boxes?

(a)

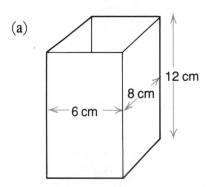

(b)

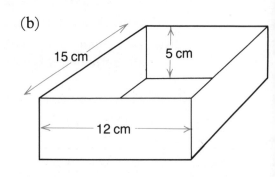

A6 This piece of metal ...   ... folds up into this tray.

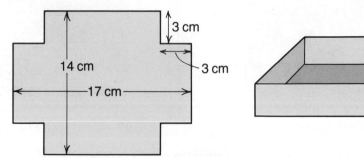

(a) Draw a rough sketch of the tray,
and mark its dimensions on your sketch.

(b) Work out the volume of the tray.

(c) What is the **area** of the metal it was made from?

This tray is 1 cm deep.

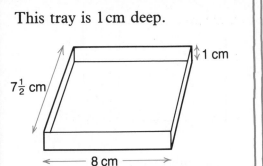

$7\frac{1}{2}$ cm

1 cm

8 cm

You can get 56 cm cubes into the tray, but you get a gap $\frac{1}{2}$cm wide.

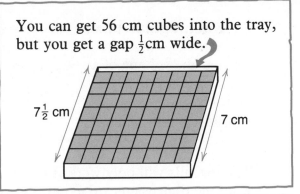

$7\frac{1}{2}$ cm

7 cm

Now you could cut 4 cubes into halves and fill the gap.

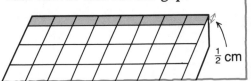

$\frac{1}{2}$ cm

So altogether you get
56 + 4 cubes into the tray.

The volume of the tray
is 60 cm^3.

A7 Work out $7\frac{1}{2} \times 8$. (Remember $7\frac{1}{2}$ and 7·5 are the same.)
What do you notice about your answer?

A8 Work out how many centimetre cubes you could get
in each of these trays. You can cut up the cubes.

(a)

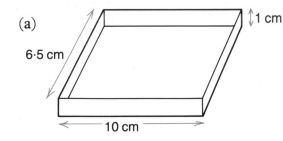

6·5 cm

1 cm

10 cm

(b)

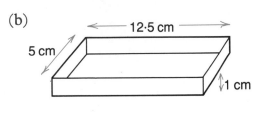

12·5 cm

5 cm

1 cm

A9 Work out how many cm cubes you could get in 1 layer
in each of these boxes.
Then work out the volume of each box.

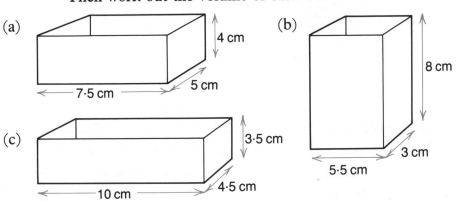

(a)

4 cm

5 cm

7·5 cm

(b)

8 cm

3 cm

5·5 cm

(c)

3·5 cm

10 cm

4·5 cm

3

B Rectangular boxes

To find the volume of a box like this you can multiply

length × *width* × *height*.

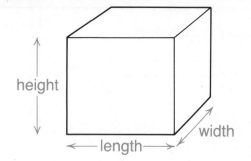

The formula will only work if the box is **rectangular,** like this one. If all the dimensions are in cm, the volume will be in cubic cm.

B1 For which of these boxes will the formula work? Write Yes or No for each part.

Then work out the volumes of the boxes for which the formula works.

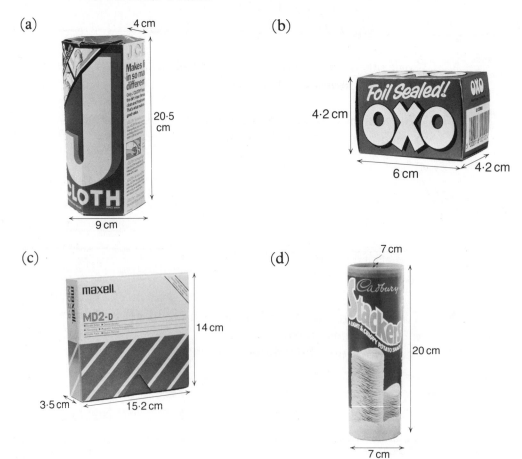

(a)   4 cm   20·5 cm   9 cm

(b)   4·2 cm   6 cm   4·2 cm

(c)   14 cm   3·5 cm   15·2 cm

(d)   7 cm   20 cm   7 cm

discussion point

How could you **estimate** the volumes of the other boxes?

c Capacity

Many drinks come in
cardboard boxes or cartons.

This is a small box
of milk chocolate drink.
The dimensions of the
outside of the box
are shown.

8·2 cm

6·4 cm   4 cm

C1 Work out the volume of the Break Time box in cm³.

C2 The Break Time box holds 200 ml of milk drink.
Could the volume of the **inside** of the box
be about 200 cm³?

A volume of 1 cubic cm is the same as 1 millilitre.

C3 A litre is 1000 ml.
  (a) How many cubic cm is 1 litre?
  (b) How many cubic cm is ½ litre?
  (c) This box holds ½ litre
      of milk.
      The dimensions of
      the outside of the
      box are marked.

      Work out the volume
      of the box.

  (d) How much bigger
      is the volume of the box
      than the milk inside?

9 cm

500 ml

9·6 cm   6 cm

10·6 cm

6·4 cm   4·1 cm

7·6 cm

7·6 cm   4·7 cm

C4 Both boxes
hold 250 ml.
Which box
has a volume
closest to
250 ml?

C5 This box holds Munch Bunch Freeze Pops!

4·2 cm

14·4 cm

20 cm

(a) What is the volume of the box?

(b) The volume of the Freeze Pops is 480 ml. About what percentage of the box do the Freeze Pops take up?

(Work out $\dfrac{\text{volume of Freeze Pops}}{\text{volume of box}}$ as a percentage.)

C6 This box holds cans of Harp Lager.

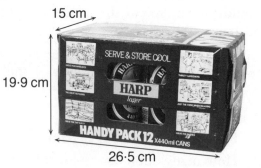

15 cm

19·9 cm

26·5 cm

(a) Roughly what is the volume of the box?

(b) Each can holds 440 ml. How many ml of lager are in the box?

(c) Roughly what percentage of the box does the lager take up?

22 cm

15·7 cm

11·3 cm

C7 This is a 3 litre wine box of Colman's red wine.

(a) Roughly what is the volume of the wine box?

(b) How many ml of wine are in the box?

(c) What percentage of the box does the wine take up?

Discuss why you think more and more drinks are being sold in boxes.

D Other units

D1 Large volumes are often measured in **cubic metres** (**cu.m.** or **m³**).

Look at this empty swimming pool.
When it is filled, the water will be 1·8 m deep everywhere (there is no deep or shallow end).

(a) Roughly what is the volume in cubic metres?

(b) If it was filled to the brim the water would be 1 m 95 cm deep.
    What would the volume of water be then?
    (You will need to write 1 m 95 cm as a decimal of a metre.)

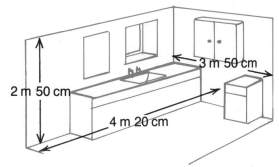

XTRACT FANS

Model	Room size	Price
XGC 6	up to 50m³	£ 37.00
XGC 9	50m³–100m³	£ 85.50
XGC 12	over 100m³	£168.75
		PLUS 18% VAT

D2 The picture shows the dimensions of a kitchen.

(a) Which Xtract fan should you fit in the kitchen?

(b) How much will it cost **including** VAT?

D3 You can buy new top soil for gardens.
It comes in skips.
A skip holds 6 m³ and costs £27.

(a) Suppose you put 10 cm of top soil on the garden in the picture.
    You can think of the top soil as a rectangular block.
    What would the volume of the top soil be?
    (Remember to work in metres.)

(b) How many skips of soil would you need?

(c) 1 m³ of top soil weighs about 2 tonnes.
    How much would the soil weigh altogether?

(d) How much would it cost?

(e) How much would it cost to put 12 cm of top soil on the garden, instead of 10 cm?

7

Review: without a calculator

Do NOT use a calculator in this review.
Try to do as much in your head as you can.

1   Kevin is checking his little sister's homework.
Here is the homework.
Check it yourself – for each part write down whether it is
right or wrong.

1. 14
  × 6
  64
	2. 13
	 +29
	  32
		3. 100
		  −17
		  93
			4. 65
			 × 2
			130

5. 25
  × 6
140
	6. 19
	 +37
	  57
		7. 87
		 −29
		  62
			8. 36
			 × 3
			118

9. 36
 × 7
252
	10. 37
	 −19
	  56
		11. 22
		 × 5
		1010
			12. 29
			 +26
			  55

2   How much will each of
these cost?

(a) 2 granary loaves

(b) 2 doughnuts and
a lardy cake

(c) 3 cream horns and
2 sausage rolls

(d) 5 Danish pastries

WHOLEMEAL LOAVES 51p
GRANARY LOAVES 55p
DOUGH NUTS 26p
LARDY CAKES 24p
CREAM HORNS 33p
ICED FANCIES 27p
DANISH PASTRIES 32p
SAUSAGE ROLLS 33p
PORK PIES 47p
CORNISH PASTIES 63p

3   What change do I get from £1 if I buy
a wholemeal loaf and an iced fancy?

4   How many pork pies could you buy for £1?

5   How many lardy cakes could you buy for £1?

To work out $\frac{1}{5}$ of something, you just divide by 5.
So $\frac{1}{5}$ of £20 is £20 ÷ 5 = £4.

6   Work out
    (a) $\frac{1}{5}$ of £30   (b) $\frac{1}{4}$ of £12   (c) $\frac{1}{5}$ of £50
    (d) $\frac{1}{3}$ of 18 cm   (e) $\frac{1}{6}$ of 24 m   (f) $\frac{1}{4}$ of £6

7   In a dog show there are 60 dogs.
    $\frac{1}{4}$ of the dogs win prizes.
    How many dogs **don't** win prizes?

8   Three friends drive 120 miles to a concert.
    Jan drives $\frac{1}{3}$ of the way. John drives $\frac{1}{4}$ of the way.
    Jim drives the rest.
    How far do they each drive?

To work out $\frac{2}{5}$ of £20, you first need to work out $\frac{1}{5}$ of £20.
$\frac{1}{5}$ of £20 is £4. $\frac{2}{5}$ is double $\frac{1}{5}$.   So $\frac{2}{5}$ of £20 is £4 × 2 = £8.

9   Work out
    (a) $\frac{3}{5}$ of £10   (b) $\frac{4}{5}$ of £10   (c) $\frac{2}{5}$ of £30
    (d) $\frac{4}{5}$ of £30   (e) $\frac{2}{5}$ of 30 cm   (f) $\frac{4}{5}$ of 60 kg

10 Work out
    (a) $\frac{2}{3}$ of £30   (b) $\frac{2}{3}$ of 18 kg   (c) $\frac{3}{4}$ of 20 cm
    (d) $\frac{3}{4}$ of 40 m   (e) $\frac{7}{10}$ of 40 km   (f) $\frac{3}{10}$ of 60 kg

11 (a) What is 10% of £1?   (b) What is 10% of £6?
    (c) What is 10% of 50p?   (d) What is 10% of £6·50?

12 Work out each of these.
    (a) 20% of £6   (b) 20% of £9   (c) 5% of £6   (d) 5% of £8

25% is the same as one-quarter.
So to work out 25% you can divide by 4.

13 Work these out.
    (a) 25% of 16 m   (b) 25% of 80 kg   (c) 25% of £10   (d) 25% of 30 cm

14 $33\frac{1}{3}$% is the same as one-third.
    (a) What is $33\frac{1}{3}$% of £60?
    (b) What will a £60 coat cost in the sale?
    (c) What will a £120 suit cost in the sale?

9

2 Maps and roads

A Main roads

The roads on this map have been used for hundreds of years.

The Romans built most of these roads about 2000 years ago.
They linked Roman army bases with London.

In the beginning, the roads only had names like Ermine Street and Watling Street.

Then Ermine Street was called the A1, Watling Street the A5, and so on.

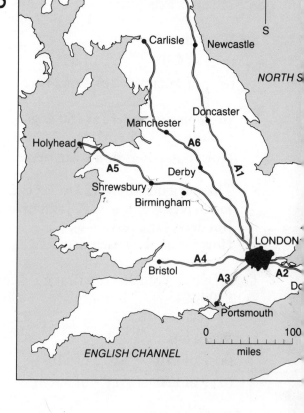

A1 The A1 joins London and Edinburgh.
Copy this table and complete it for all the roads on the map.

Road	Joins London an...
A1	Edinburgh
A2	Dover
A3	

A2 One big city on the map is not on any of these roads.
Which city is that?

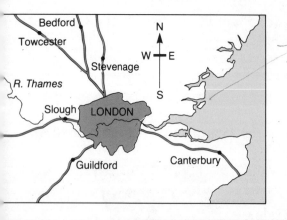

A3 This map is an enlargement of the area around London.
It shows these main roads near London.

Make a sketch of this map.
Put in the road numbers.

A4 (a) About how far is it from London to Shrewsbury along the A5?

(b) A Roman army could march about 20 miles a day. How long would this journey take them?

10

discussion points

This map shows some
of the motorways in
England and Wales.

Compare them with
the Roman roads.

The motorways do not
usually go through
the centre of a town.
Why is that?

Does the M1 go between
the same towns as the A1?

What about the
other motorways –
do they go to the same
towns as the Roman roads?

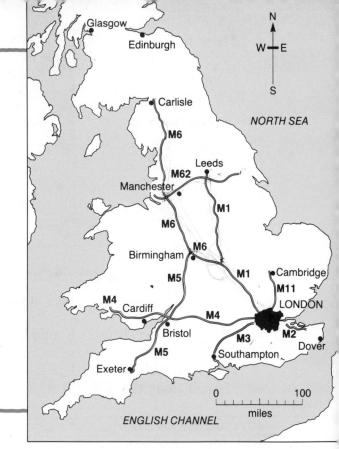

A5 This map shows the motorway
junction near Bristol.
Find it on the large map.

(a) What is the number
of motorway A?

(b) Which is motorway B?

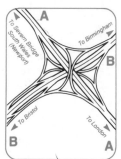

A6 Here are two more motorway
junctions.
Work out where they must be on the large map.
Write down the numbers of motorways A, B, C, and so on.

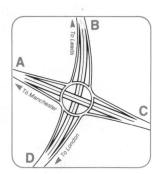

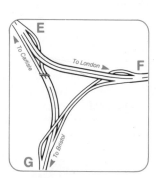

B How far?

This map shows part of Kent, the A2 and A20.

The numbers in red are the distances between the towns in miles.

From the middle of London to Chatham is 32 miles.

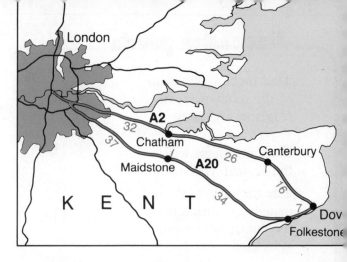

B1 (a) How far is it from Chatham to Canterbury?

(b) How far is it from Canterbury to Dover?

B2 How far is it from London to Dover along the A2?

B3 How far is it from London to Dover if you go along the A20, through Folkestone?

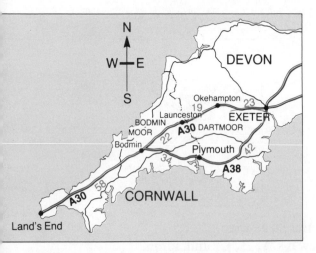

B4 This map shows part of Devon and Cornwall.

How far is it from Land's End to

(a) Bodmin

(b) Plymouth

(c) Okehampton

B5 There are two roads from Exeter to Bodmin. Which is shorter?

B6 This is a page from a guidebook to Devon and Cornwall.

Can you work out the missing words?

For example, the first word is *west* or *east*.

Copy the page and fill in the missing words.

Exeter – Land's End

As you leave Exeter, look for signs to the A30.
The road goes ▨▨▨ After a few miles you will see
Dartmoor on the ▨▨ hand side. The first town
you come to is ▨▨▨. It is ▨ miles from Exeter.
Between ▨▨▨ and ▨▨▨▨ you will cross the
county boundary, and pass from Devon into
▨▨▨▨▨ Soon the road takes you over
▨▨▨▨. Moor.
Just before you reach Bodmin, you meet the
A▨ from ▨▨▨▨ You continue on the A ▨
Bodmin is ▨▨ than halfway from Exeter to
Land's End. ▨

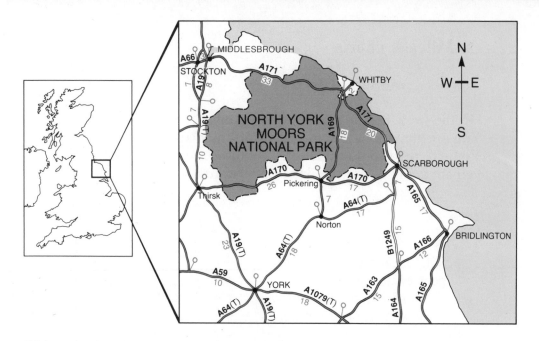

B7 This map shows part of North Yorkshire.
The distances between the red 'lollipops' – 🛈 – are marked in miles.

Find York on the map, and the **A64**(T) to Scarborough.
(The T means it is a Trunk road – a large main road.)

(a) In which compass direction does the road go to Scarborough?

(b) How far is it altogether from York to Scarborough?

B8 (a) Which other **Trunk** roads go through York?

(b) Look at the roads from York to Bridlington.
List the road numbers of the shortest route
from York to Bridlington.

(c) Which is closer to York, Scarborough or Bridlington?

B9 York is 209 miles from London.

(a) How far do you think Thirsk is from London?

(b) How far is Middlesbrough from London?

B10 Pickering is a good centre for visiting the National Park.
Make a list of 4 or 5 of the large towns on the map.
Work out the distance of each of them from Pickering.

B11 Suppose you had to organise a cycle race.
The race must start and end at York.
It has to be about 100 miles long.

Write down the route you would choose for the race.
List the road numbers and towns on the route.

c Mileage charts

You need worksheet G8–1.

On the worksheet is a map of East Anglia.
You are going to add some details to the map.
You will need to do this neatly because
you are going to use the details afterwards.

> This means that the road
> goes on to Sleaford.
> Sleaford is too far away
> to be shown on the map.

Sleaford

Peterborough

A47

C1 Some of the road numbers
are shown on the map.
Write in the other road numbers.
You can get them from this table.

Road	Towns on the road
A17	Sleaford, King's Lynn
A10	London, Cambridge, King's Lynn
A11	London, Norwich
A12	London, Ipswich, Great Yarmouth
A45	Northampton, Cambridge, Ipswich
A47	Peterborough, Great Yarmouth

C2 What are the road numbers of the roads
you would travel on to get to Great Yarmouth
from Cambridge?

C3 In which town might
you see this sign?

C4 What do you think
the missing road
numbers are on
the sign?

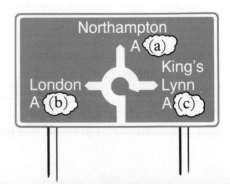

14

On the worksheet there is a mileage chart. It shows the distances between the towns on the map, using the roads shown.

For example, the chart shows that it is 10 miles from Great Yarmouth to Lowestoft.

You need a coloured pen or pencil.

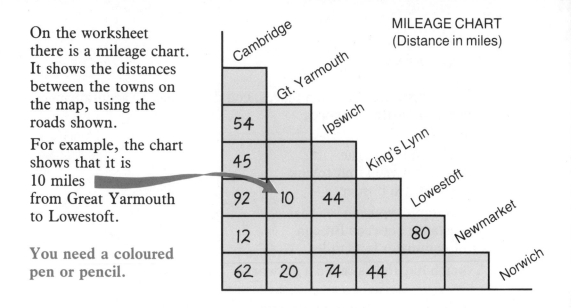

MILEAGE CHART
(Distance in miles)

C5 In colour, mark the distance from Great Yarmouth to Lowestoft on the map on your worksheet.

C6 (a) How far is it from Ipswich to Lowestoft?
   (b) Mark this distance on the map.

C7 (a) Mark the distance from Cambridge to Newmarket on the map.
   (b) The distance from Newmarket to Norwich is not in the table.
      But the distance from Cambridge to Norwich is.
      How far must it be from Newmarket to Norwich?
      Mark the distance on the map.

C8 Mark the distances between each of the towns on the map.
   You will have to do a bit of detective work!

C9 Look at the roads from Cambridge to Great Yarmouth.
   You could go via Norwich, or via Ipswich.
   (a) How far is it via Norwich?
   (b) How far is it via Ipswich?
   (c) Fill in the shorter distance in your mileage chart.

C10 Work out the shortest distances between each of the towns in the mileage chart.
   Fill each distance in.

15

D Networks

To show distances between places, you don't need an accurate map. You can show the distances on a sketch.

This sketch shows the distances between four towns near the Welsh border.

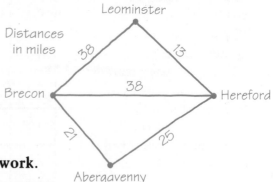

For example, you can see that the distance between Brecon and Leominster is 38 miles.

A sketch like this is called a **network**.

D1 How far is it from Brecon to Abergavenny?

D2 (a) How far is it from Brecon to Hereford?

(b) How far is it from Brecon to Hereford if you go through Abergavenny?

D3 This network shows the distances between five towns in the Midlands.

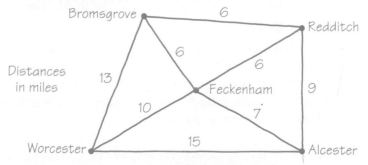

(a) How far is it from Worcester to Redditch if you go through Feckenham?

(b) How far is it through Bromsgrove?

D4 Jane wants to visit all five towns. She needs to start and end in Worcester.

One way she could do it is like this.

(a) Copy the sketch map. Mark the distances between each of the towns.

(b) How far is Jane's route altogether?

D5 Jane can visit the towns in a different order.

Worcester → Bromsgrove → Redditch → Feckenham → Alcester → Worcester

     (a) Draw a sketch map for this route.

     (b) Mark the distances between the towns on your sketch.

     (c) How far is this route altogether from Worcester back to Worcester?

D6 Look at Jane's routes in questions D4 and D5.
Can you find a route through all the towns
that is shorter than these two routes?
Make a sketch of your shorter route.

The shortest route is not always the quickest.
This network shows the **time** it takes between the five towns.

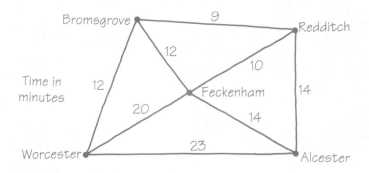

There is a motorway between Worcester and Bromsgrove
so that route is fairly quick.

D7 (a) How long does it take from Worcester to Redditch
via Feckenham?

     (b) How long does it take via Bromsgrove?

D8 Look at Jane's route between the five towns in question D4.

     (a) Make a copy of the route.
Mark the times between the towns on your copy.

     (b) How long would it take Jane to travel this route?

D9 Can you find Jane a quicker route between the five towns,
starting and finishing in Worcester?
Draw a sketch of your quicker route.

Work out how long your quicker route takes.

3 Large numbers

discussion point

When you open a newspaper you always see lots of numbers.
And you always hear lots of numbers on the radio and TV.

Look at the numbers in the headlines above.
Do you know how to say all the numbers?
Is there more than one way to say some of the numbers?

Sometimes you can write the same number in different ways.
Could you write any of the numbers above in a different way?

A Say it with numbers

Many people like reading a newspaper. The people
who write newspapers try to make the words easy to read.
But sometimes the numbers are more difficult to read than the words!

When you read a headline like this...	...the **000** tells you 'thousands'.	So you read 'ninety-two thousand pounds'.

A1 Write down what you read for each of these.

(a)

(b) **£250,000 JACKPOT!**

(c) **£105,000 SCOOP**

A2 You write 'twenty thousand' in figures as 20,000
Write each of these in figures.

(a) Sixty thousand

(b) One hundred and twenty thousand

(c) Two hundred and five thousand pounds

(d) Three hundred and sixty-three thousand pounds

A3 Half a thousand in figures is 500.
So one and a half thousand is 1500.
Write these in figures.

(a) Two and a half thousand

(b) Seven and a half thousand

(c) Ten and a half thousand

Writing numbers

You can write a number like twenty-five thousand in different ways.

You can write it using commas. 25,000

You can write it using spaces. 25 000

You can write it without either! 25000

To make it easy to read, it is best to use **commas**
when you write large numbers.

B Millions

When you read this headline
you could say that
a thousand thousand pounds
was paid out.

£1 000 000 PAY-OUT

Birmingham's top officials collected
over £1 million in wages last year.
Top pay-out was to Chief executive
Sarah C~~~~ who collected well
over

But there is a special word for a thousand thousand.
We call a thousand thousand a **million**.

Newspapers sometimes write £1m or £1million for £1,000,000.
They often write £10m for £10,000,000
    or £25million for £25,000,000 and so on.

B1 Write these headlines using *m* or *millions*
The first is done for you.

(a) **£30,000,000 THEFT**

B1(a) £30 m theft.

(b) **£5,000,000 CLAIM**

ROCK STAR
SUES MANAGER
FOR
£ 5000 000

(c) **£60,000,000 FRAUD**

B2 Write these headlines out in full.
The first is done for you.

(a) **3 million on march**

B2(a) 3,000,000 on march.

(b) **£20m profit by BR**

(c) **£100 million losses by APT**

(d) **£500m Pension fiddle**

B3 $\frac{1}{2}$ million is the same as 500,000.
So $2\frac{1}{2}$ million written out in full is 2,500,000.

Write these numbers out in full.

(a) $3\frac{1}{2}$ million   (b) $1\frac{1}{2}$ million

(c) $10\frac{1}{2}$ million   (d) $7\frac{1}{2}$ million

Calculators don't use spaces or commas
when they show a number.
When a calculator shows a million,
it usually does it like this.

The number is much clearer if you
write it using commas.

Now you can easily see the number
is one million.

1,000,000

B4   Write each of these calculator displays using commas.

(a)

(b)

B5   Write **in words** the number each calculator is showing.

(a) 6000000   (c) 10000000

(b) 200000   (d) 200000000

B6   Draw a calculator face showing half a million.

B7   The number on this calculator
is about 2 million.

2167013

Match each of the numbers below with its calculator face.

(a) 40 million

(b) 4 million

(c) 400 thousand

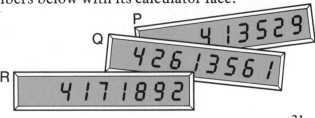

P   413529

Q   4261356 1

R   4171892

C Rounding

When you read or write numbers
you often don't need to be exact.

7241 is between 7000 and 8000
but it is nearer to 7000.

Eugen has rounded 7241 **to the
nearest thousand**.

7000   8000

7241

C1 Write these headlines to the nearest thousand.

(a)
3340 JOBS TO GO

(b)
£8746 Robbery

(c)
42,346 watch match

(d)
£27,898 BONUS!

C2 Round these headlines to the nearest million.

(a)
£5,113,279 WIN!

(b)
£8,213,467 wage bill!

(c)
£10,367,129 loss

(d)
£1,496,317 PROFIT!

C3 Write each of these numbers to the nearest hundred.
(a) 735   (b) 498   (c) 5247   (d) 7841

C4 Is the number on this
calculator closest to

1963478

(a) 1 million   (b) 1½ million   (c) 2 million

C5 2497376 is 2½ million to **the nearest ½ million**.
Write each of these to the nearest ½ million.

(a) 7386497 (Is it nearer to 7 million or 7½ million?)

(b) 8612379   (c) 10901306

22

D Significant figures

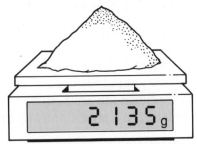

When you see a number like 2135...

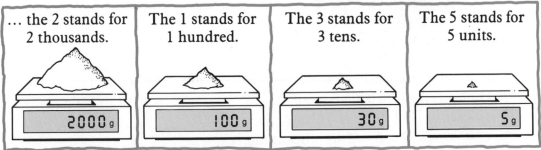

| ... the 2 stands for 2 thousands. | The 1 stands for 1 hundred. | The 3 stands for 3 tens. | The 5 stands for 5 units. |

In 2135, the 2 is the **most significant** figure.
It stands for the largest number.

You can round 2135 in various ways.

| 2135 is **2000** to the nearest **thousand**. | 2135 is **2100** to the nearest **hundred**. | 2135 is **2140** to the nearest **ten**. |

When you round 2135 to the nearest thousand we call it
rounding to 1 significant figure.

You can round other numbers to 1 significant figure.

In 391 the most significant figure is the 3.
It stands for 3 hundreds.
To round 391 to 1 significant figure
we need to round to the nearest hundred.

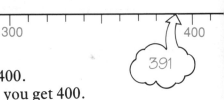

391 is between 300 and 400. But it is closer to 400.
So when you round 391 to 1 significant figure, you get 400.

D1 Round each of these numbers to 1 significant figure.
  (a) 476   (b) 5264   (c) 7835
  (d) 971   (e) 4097   (f) 249

D2 (a) In 48 631, which is the most significant figure?
  (b) What number does the most significant figure stand for?
  (c) Is 48 631 closer to 40 000 or 50 000?
  (d) Round 48 631 to 1 significant figure.

23

D3 Round each of these numbers to 1 significant figure.

(a) 24135   (b) 38156   (c) 89236   (d) 10387

D4 Write a headline for this newspaper story.
Include the amount Mrs Taylor won, to 1 significant figure.

> Mrs Monica Taylor really hit the jackpot when she won £78352 in Newtown's latest Lottery. Monica, a cleaner at Newtown School, told our reporter "I just couldn't believe it. I've never...

You can round numbers to 2 significant figures.
Mrs Taylor won £78352 in the lottery.
The two most significant figures in 78352 are the 7 and the 8.

78352 is between 78000 and 79000.
It is closer to 78000.

So £78352 is £78000 to 2 significant figures.

78000   79000

78352

D5 Round each of these numbers to 2 significant figures.

(a) 4176   (b) 32364   (c) 478   (d) 4932

(e) 82476   (f) 87921   (g) 3486   (h) 59810

D6 In this headline, the price is rounded to 2 significant figures.

Round each of these headlines to 2 significant figures.

> **£37,000,000 VASE!**
> A vase fetched £36850000 on Monday – a record for a piece of pottery in the UK. The anonymous

(a) **£17,230,000 RANSOM**

(b) **£77,950,000 DEAL**

(c) **387,942 more jobs**

(d) **1,867,000 see Pope**

Have a go

4 Continuous data

HIT THE LINE!   A GAME FOR TWO PLAYERS

You need: a sheet of graph paper,
a ruler, a pencil and a 2p coin each.

Take the piece of graph paper.
Cut it in half across the middle.

Take half a piece each.

You each draw a line about 20 mm
from the end of your graph paper.
Label it Target line

Label each piece of graph paper
every 10 mm up to 160 mm.
Start with 0 mm on the Target line

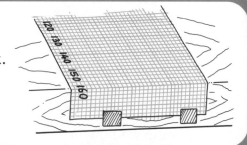

Fold each piece over on
the 160 mm line.
Put the fold over the edge of your desk.
Tape it into position.

You are now ready to play
        HIT THE LINE!

You each play on your own
sheet of paper.

Put the 2p coin over the edge
of the graph paper.
Tap it up the paper.
Try to get as close to the
Target line as you can.

If the coin goes over the line
have another go.

When the 2p stops, look at
the edge nearest the

Hold the 2p firmly, and make
a **small mark** against the edge.

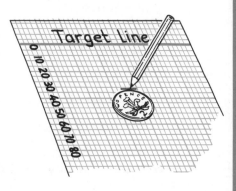

Tap the coin and mark the
edge 50 times.

Now you have to decide who won!
You could just say the person with the mark
nearest to the Target line wins.

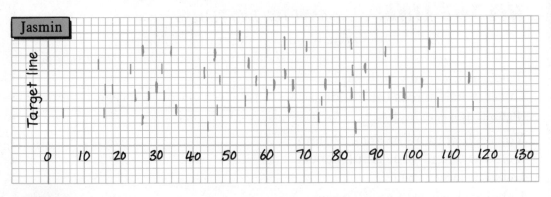

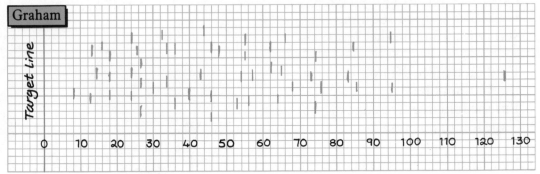

So here Jasmin would win.
But Graham seems to be better than Jasmin.

To see how good a player is, you can draw a bar chart.
You can do this on the graph paper you used for the game.
We'll see how to do this for Jasmin before you do your own.

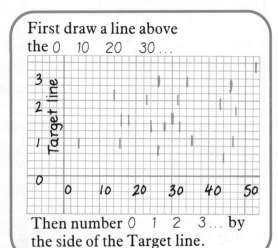

First draw a line above
the 0   10   20   30...

Then number 0   1   2   3... by
the side of the Target line.

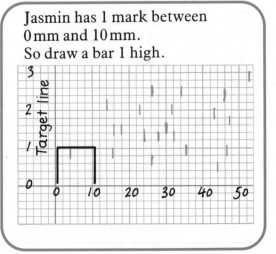

Jasmin has 1 mark between
0mm and 10mm.
So draw a bar 1 high.

Jasmin has 4 marks between 10 mm and 20 mm.
So draw a bar 4 high.

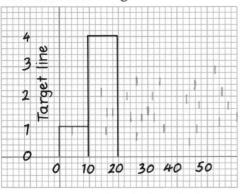

Jasmin has 5 marks between 20 mm and 30 mm.
But she also has one mark at **exactly** 30 mm.

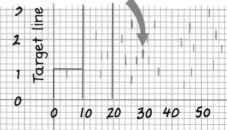

Should this count in the 20 to 30 bar or in the 30 to 40 bar?

As long as we always do the same it doesn't really matter.

Where a mark is on the edge of a bar, count it in the next bar up.

So we will count 30 mm in the 30 to 40 bar.

We will leave the 20 to 30 bar with just 5 marks in it.

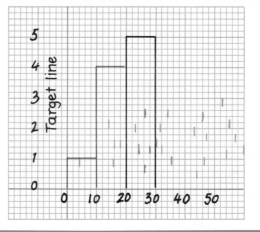

A1 Jasmin also has a mark at exactly 60 mm.
    Will this count in the 50 to 60 bar, or in the 60 to 70 bar?

A2 Jasmin has a mark at 80 mm.
    Which bar will this count in?

A3 (a) On a new sheet of graph paper, draw Jasmin's bar chart.
        You don't need to copy her marks!
    (b) Draw Graham's bar chart.

A4 (a) Look at the charts you have drawn for Jasmin and Graham.
        Decide on a fair rule to tell you who is the best player.
    (b) Now draw the bar chart for your own game.
        Compare your bar chart with your friend's chart.
        Use your fair rule to decide who won the game.

B Frequency

This shows Theo's marks when he played Hit the line!

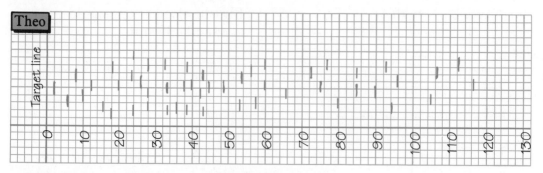

Theo did not want to draw bars 10 mm wide.
He wanted to group the marks in bars 20 mm wide.
He decided to draw up a table first.

The first group must go from 0 mm to 20 mm.
Remember a mark of 20 mm **doesn't** go in this group.
So we label it *0 mm and less than 20 mm*.

Marks	
0 mm and less than 20 mm	

The next group goes from 20 mm to 40 mm.
A mark of 20 mm **does** go in this group.
A mark of 40 mm **doesn't** go in this group.
We label it *20 mm and less than 40 mm*.

This column shows how many marks there are in each group.
We label it *frequency*.

Marks	Frequency
0 mm and less than 20 mm	
20 mm and less than 40 mm	

B1 (a) Copy Theo's table and complete it.
Count his marks in each group. Fill in the Frequency column.

(b) On graph paper draw a bar chart for Theo's marks.
Label across the graph like this.

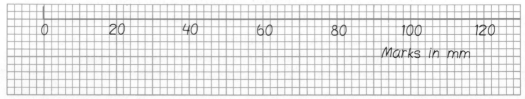

We call tables like the one Theo used **frequency tables**.
The bar chart showing the frequencies is called a **frequency chart**.

Here is a different frequency chart.
It shows the length of some
worms found in a clay soil.

For example, you can see that
3 worms were between 0 and 2
centimetres long.
They were less than 2 cm long.

A worm that was exactly
2 cm long would count
in the 2 to 4 column.

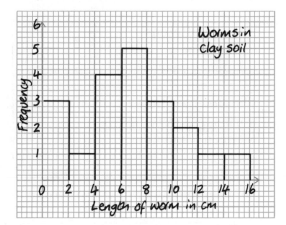

B2   (a) How many worms were between 2 and 4 cm long?
     (b) How many worms altogether were less than 4 cm long?
     (c) How many worms were less than 8 cm long?
     (d) How many worms were measured altogether?

B3   This frequency chart
     shows the lengths of
     worms in a sandy soil.

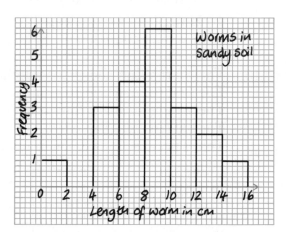

     (a) How many worms were less than 2 cm long?
     (b) How many altogether were less than 8 cm long?
     (c) How many worms were measured here altogether?
     (d) Do you think that the worms were longer on average
          in the clay soil or in the sandy soil?

29

Here is a frequency chart for worms found in a marsh.

It looks different to the other worm frequency charts.

The first bar shows worms that were between 0 and 4 cm long.

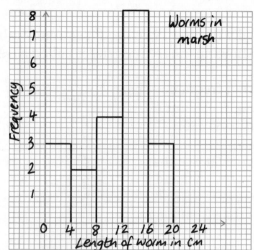

For example, if a worm was 3·6 cm long, it would count in the first bar.

B4 A worm 4 cm long counts in the 4 to 8 cm column.
    (a) In which bar does a worm 8 cm long count?
    (b) In which bar does a worm 16 cm long count?

B5 (a) How many worms in the marsh frequency chart
        were less than 12 cm long?
    (b) How many were at least 12 cm long, but
        less than 20 cm long?

B6 The chart below shows the weights of snails found in a garden.
The first bar starts at 10 grams.
There were no snails found that weighed less than 10 g.
The bars in this chart are wider than before.
That's because the person drawing them thought they looked better.

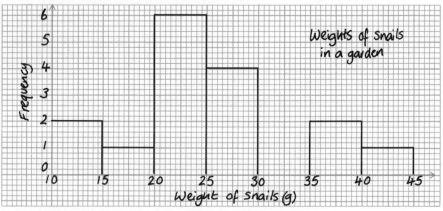

    (a) How many snails weighed 35 grams or more?
    (b) Without doing any working out, **roughly** what would you say
        is the average weight of snails in the garden?

c Grouping

Jan's class measured the weights of 30 young rabbits in grams.

113	45	126	120	58	100	96	177	197	42
168	162	87	108	115	49	84	76	71	185
114	175	101	159	181	117	83	62	122	81

Each person in the class draws a frequency chart of the weights.
They each decide how to group the weights.

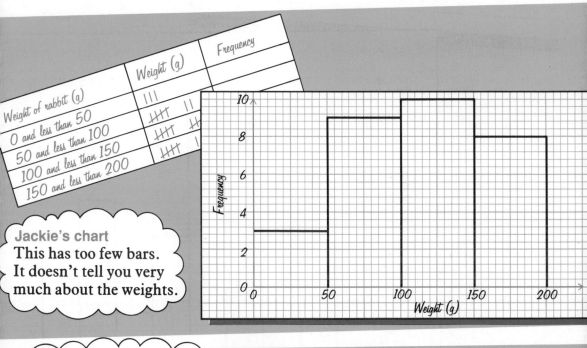

Weight of rabbit (g)	Frequency
0 and less than 50	|||
50 and less than 100	⊞⊞ ||
100 and less than 150	⊞⊞ ⊞⊞
150 and less than 200	⊞⊞	

Jackie's chart
This has too few bars.
It doesn't tell you very
much about the weights.

Brian's chart
This has too many bars!
It is very squashed.
Also there are no weights
less than 40 grams.
So you don't need to start at 0.

Weight of rabbit (g)	Tally	Freque
0 and less than 10		0
10 and less than 20		0
20 and less than 30		0
30 and less than 40		0
40 and less than 50	|||	3
50 and less than 60			1
			1
				2
						4
			1
					3
						4
					3
		0

Weights of
rabbits

31

Weight (g)	Tally	Frequency
40 and less than 60	IIII	4
60 and less than 80	III	3
80 and less than 100	HHT	5
100 and less than 120	HHT II	7
120 and less than 140	III	3
140 and less than 160	I	1
160 and less than 180	IIII	4
180 and less than 200	III	3

Jan's chart
This is about right.
The chart starts at 40, which is sensible.
It has about the right number of bars so you can see how the weights vary.

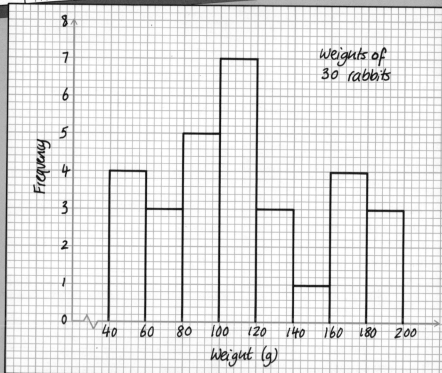

Weights of 30 rabbits

Drawing a frequency chart

1 Look through all the data you need to chart.
What is the smallest value of the data?
What is the largest value?

2 Decide how to group the data with a sensible number of bars.

3 Draw up a frequency table and tally the data.
Fill in the *frequency* column.

4 Draw the axes for your chart and label them.

5 Now you can draw the chart and label it.

C1 Saed has planted some trees.
After one year he measures the heights of 30 of the young trees.
Here are the heights in centimetres.

179	75	140	37	65	189	155	99	57	160
77	86	58	155	119	181	73	84	176	39
40	155	160	77	172	131	160	71	175	156

(a) Check that the smallest tree is 37 cm high.

(b) What is the height of the tallest tree?

(c) Copy and complete this frequency table.

Height of tree (cm)	Tally	Frequency
20 and less than 40		
40 and less than 60		

(d) Draw a frequency chart for the height of the trees.

C2 Sarah is researching into bird-song.
She times the length of song of thrushes.
Here are her timings in seconds for 40 songs.

148	167	399	214	223	487	243	400	265	271
287	300	312	345	321	245	323	299	267	378
398	371	543	298	256	361	249	412	200	281
340	365	271	106	376	370	435	200	198	342

(a) Check that the shortest song lasted 106 seconds.

(b) How long was the longest song?

(c) We want to group the songs into **roughly** 10 equal groups.
What would be a sensible choice for
the length of each group in seconds?

(d) Draw up a frequency table.
Start your first group   100 seconds and less than . . .

(e) Complete the frequency table and draw a frequency chart.

C3 25 people were shown a line exactly 30 cm long.
They were asked to estimate how long the line was.
Here are their estimates in centimetres.

| 32·5 | 25·5 | 32 | 27·5 | 34 | 30·5 | 30·5 | 38 | 31 | 41·5 | 33 | 39 | 32·5 |
| 32·7 | 34·4 | 28 | 30 | 36·5 | 27 | 36·5 | 21 | 36·5 | 37 | 24·5 | 35 | |

(a) Choose sensible groups and draw up a frequency table.

(b) Fill in your frequency table.

(c) Draw a frequency chart for the estimates.

Review: decimals, percentages and fractions

This scale shows decimals and percentages.

10% is the same as 0·1, 24% is the same as 0·24, 5% is the same as 0·05.

1   Write these percentages as decimals.
     Try to do them without using the scale.
     (a) 40%   (b) 45%   (c) 90%   (d) 9%
     (e) 30%   (f) 3%   (g) 19%   (h) 7%

It is easy to work out 20% of £4 in your head.

10% of £4 is 40p ...   ... so 20% of £4 is 80p.

It is not so easy to work out 23% of £4 in your head!
But you can do it like this with a calculator.

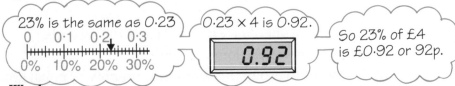

23% is the same as 0·23   0·23 × 4 is 0·92.

0.92

So 23% of £4 is £0·92 or 92p.

2   Work out
     (a) 18% of £9   (b) 24% of £9   (c) 80% of £9   (d) 8% of £9

3   Jake and Irma work in a small café.
     Jake earns £74 a week. Irma earns £82·50 a week.
     They both get a pay-rise of 6%.
     How much extra will they each get?

WERE
£34

4   In a sale, 18% is taken off all prices.
     (a) How much is taken off these shoes?
     (b) How much do the shoes cost in the sale?

5   The old Squelch bar weighed 120 grams.
     The new one is 18% heavier.
     How much **extra** do you get in the new bar?

EVEN SQUELCHIER   18% MOR

It is easy to work out $\frac{3}{4}$ of £20.
$\frac{1}{4}$ of £20 is £5.
$\frac{3}{4}$ of £20 is 3 times as big, so $\frac{3}{4}$ of £20 is £15.

It is not so easy to work out $\frac{3}{4}$ of £17·60 in your head!
But you can do it with a calculator.

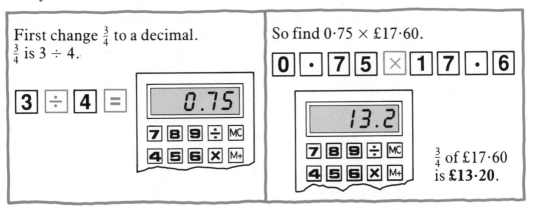

First change $\frac{3}{4}$ to a decimal.
$\frac{3}{4}$ is 3 ÷ 4.

So find 0·75 × £17·60.

$\frac{3}{4}$ of £17·60 is **£13·20**.

6   Use a calculator to change these fractions to decimals.
   (a) $\frac{1}{5}$ (b) $\frac{3}{5}$ (c) $\frac{1}{8}$ (d) $\frac{3}{8}$ (e) $\frac{17}{20}$ (f) $\frac{3}{16}$

7   Work out
   (a) $\frac{3}{4}$ of £94·60   (b) $\frac{3}{4}$ of £27·20
   (c) $\frac{7}{20}$ of £18·40   (d) $\frac{9}{20}$ of £6·20
   (e) $\frac{2}{5}$ of 830 km   (f) $\frac{3}{8}$ of 180 kg

8   Cathy is planning a new garden.
   Altogether, the garden covers 720 m².
   (a) $\frac{2}{5}$ of the garden will be lawn.
       How many m² is that?
   (b) $\frac{3}{8}$ of the garden will be used
       to grow vegetables.
       How many m² is that?
   (c) A pond will take up $\frac{3}{20}$ of the garden.
       How many m² will the pond cover?

9   New House School has 840 pupils.
   (a) $\frac{3}{4}$ of the pupils learn Spanish. How many is that?
   (b) $\frac{2}{5}$ of the pupils are in the school choir. How many is that?
   (c) $\frac{5}{8}$ of the school play hockey. How many **don't** play hockey?

10 1344 people go to watch the Royal Ballet.
   $\frac{6}{7}$ are girls. How many is that?

35

5 Cuboids

A Nets of a cube and a cuboid

This is a drawing of a cube.
In the drawing you can see 9 of its edges, and 7 of its corners.
The corners are called **vertices**.

There are 3 more edges and one more **vertex** round the back of the cube. The hidden edges can be shown by dotted lines, like this.

A cube has 6 square **faces**.
Imagine the cube to be made of card. The 6 faces can be 'opened out' by cutting along edges, like this.

When the card is flattened out, it looks like this.

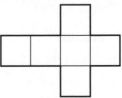

This is called a **net** of the cube.

This is also a net of a cube.

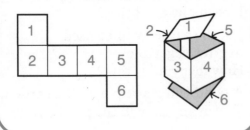

But this is **not** a net of a cube.

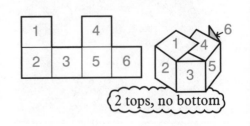

2 tops, no bottom

A1 Is this a net of a cube?
Draw it and cut it out if you are not sure.

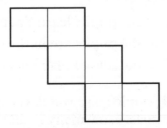

You need worksheet G8–2.

A2 Look at the diagrams at the top of the worksheet.
Which of the diagrams are the nets of a cube?

Try to answer first without cutting out the shapes.
Then cut out the shapes and fold them up to check your answers.

If you make a cube from paper or card, you need
'tabs' or 'flaps' to glue the paper or card together.

A 'net' does not include the tabs.
These are something extra added to the net.

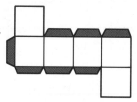

Every face of a cube is a square.
If some of the faces were not squares
it would not be a cube.

A solid with every face a rectangle
is called a **cuboid**.

A cube is a special sort of cuboid, just like
a square is a special sort of rectangle.

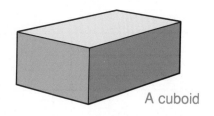

A cuboid

This cuboid is 2 cm by 3 cm by 4 cm ...   ... and here is a net for it.

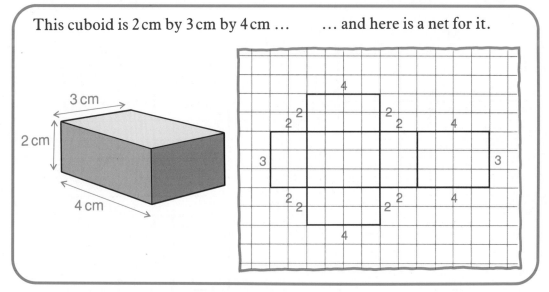

A3 (a) Draw a net for this cuboid
on worksheet G8–2.
The base has already
been drawn for you.

(b) Cut out your net and
fold it up to check that
it is correct.

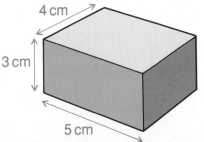

37

A4 Are these nets of cuboids? Try to answer without making them.

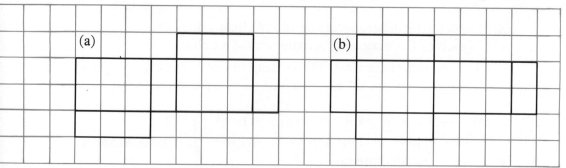

(a)

(b)

A5 Which of the drawings below are nets of cuboids?
Answer first without drawing or making them.
Then check any you are unsure about by making them.

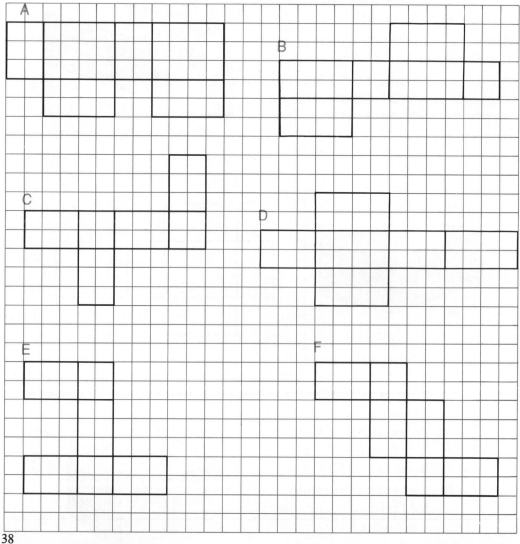

A

B

C

D

E

F

Cardboard boxes are often in the shape of a cuboid.
They are made flat and then folded.

This is an 'Oxo' box flattened out, full-size.
The glue for sticking it together goes in one place only, the
white space on the flap at the bottom.

A6 (a) Measure the flattened box to
       find the length, width and
       height of the box.

    (b) Work out the volume of the box.

A7 Look at the cut-out at the top of the worksheet.
You can make this box from it.

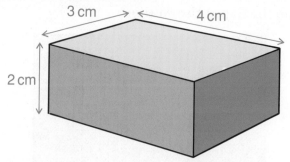

3 cm   4 cm

2 cm

(a) On the worksheet, two faces of the box are labelled A.
This is the pair of faces with the biggest area.

Find the pair of faces with the next biggest area.
Label this pair B.

Label the pair with the smallest area C.

(b) Cut along all the solid lines of the cut-out.
Fold along all the dotted lines.

Stick the box together.
Make sure that the letters are on the outside.

Keep your box. You will need it in the next chapter.

A8 Look at this box.

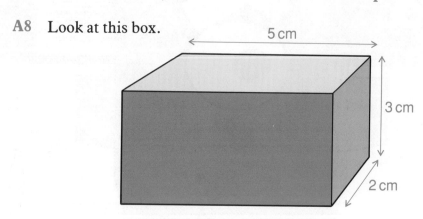

5 cm

3 cm

2 cm

(a) On the bottom of the worksheet, draw a net for the box.
Then draw tabs (you need 7 altogether).

(b) Label the faces like you did in question A7.
Cut out your net with its tabs.
Fold and stick the box together.

6 Probability

A Is it likely?

... a few light showers are likely over southern England and there's a slim chance of thundery rain in places. Overnight there may be a touch of frost on high ground with a good chance of snow in sheltered spots....

Weather forecasters try to say what will happen in the future.
They try to **predict** the weather.

They use words like 'fairly certain', 'a good chance',
'very likely', 'a faint possibility'.

Words like these describe **how likely** it is that something will happen.

You can use a scale like this to show how likely things are.

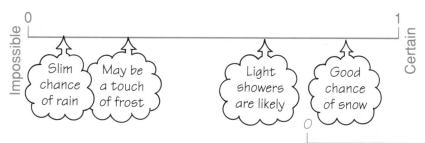

A1 Draw a line 10 cm long.
Put these weather forecasts on it.

(a) Thunderstorms are almost certain.

(b) There is no chance at all of a hurricane.

(c) It is very likely to be warm and sunny.

(d) There is a fifty-fifty chance of fog in the morning.

A2 Here are the rules of a simple game.
You roll two dice.
The result is the **difference** between
the scores on the two dice.

So here the result is 5 − 3 = 2.

Copy this table.

Result	How likely
0	
1	
2	
3	
4	
5	
6	

Without playing the game,
think about how likely
each result is.

Then fill in the How likely column.
Use words like fairly likely
most likely   least likely and so on.

A3 **You need a partner and two dice.**
Both of you make a copy of this table.

Result	Tally	Frequency	%
0			
1			
2			
3			
4			
5			
6			

One person throws the dice, the other person
puts tally marks in their copy of the table.

Throw the dice 50 times.
The result is the **difference** between the scores on the dice.

When you have thrown the dice 50 times
the person filling in the tally marks fills in the frequency column.

Then swap over.
Throw the dice 50 times and fill the other person's table in.

A4 Now work out your % column.
Suppose you got a result of 0
8 times out of 50.

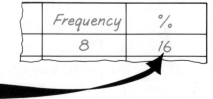

Frequency	%
8	16

8 times out of 50 is the same as $\frac{8}{50}$.
$\frac{8}{50}$ as a decimal is $8 \div 50 = 0.16$, or 16%.

Now work out your own % column and fill it in.

A5 Draw a line 10cm long.
Mark it every centimetre, and label it like this.

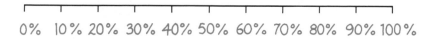

0%   10% 20% 30% 40% 50% 60% 70% 80% 90% 100%

Show each of your results on
the line with an arrow, like this.
(Use your own percentages!)

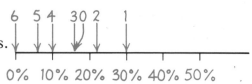

0% 10% 20% 30% 40% 50%

Look at the number of times you got a result of 0.
You and your partner probably have a different percentage of 0s.
But if you kept on throwing the dice,
the percentage of 0s would eventually settle down.

Unless there is something wrong with the dice
the percentage of 0s would settle down to about 17%.

We call this the **probability** of getting a result of 0.

This means that if you threw the dice 200 times (for example)
you would expect a result of 0 about 17% of 200 times,
about 34 times.

You would **not** expect **exactly** 34 results of 0,
but you would expect **about** 34.

A6 This table shows the probabilities of the different results.

Result	0	1	2	3	4	5	6
Probability	17%	28%	22%	17%	11%	5%	0%

If you played the game 200 times
(a) roughly how many results of 1 would you expect?
(b) roughly how many results of 2 would you expect?
(c) would you be surprised if a result of 3
came up 70 times?
(d) would you be surprised if you got a result of 4
20 times?

A7 Write down what
this means
in a simpler way.

The probability
of getting a result
of 6 is 0%.

B Percentages

You can show probabilities on a probability scale.
The scale goes from 0% for something that is impossible
to 100% for something that is certain to happen.

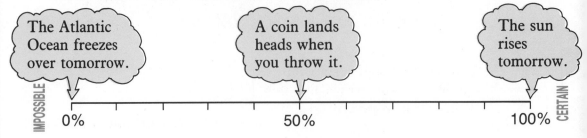

The Atlantic Ocean freezes over tomorrow.

A coin lands heads when you throw it.

The sun rises tomorrow.

IMPOSSIBLE
0%   50%   100%   CERTAIN

Other things have probabilities between 0% and 100%.

When you throw a coin, it might land heads or tails.
Unless you have an unfair coin, the probability
it lands heads is 50%.

If you threw the coin 20 times, you would expect
about 50% of the 20 throws to come up heads.

B1 The floor of a room is covered with carpet tiles.

$\frac{1}{4}$ of the tiles are red.

$\frac{3}{4}$ of the tiles are grey.

Suppose you throw a drawing pin onto the floor.

(a) What is the probability it will fall on a red tile?

(b) What is the probability it will fall on a grey tile?

B2 This means that the person thinks the probability of a dry summer is about $\frac{1}{10}$.

What is that as a percentage?

The chances of a dry summer are about 1 in 10.

discussion points

Is the probability that the sun rises tomorrow **exactly** 100%?

Can you think of other things with a probability of 100%?

What other things have a probability of 0%?

'Your chances of winning the pools are 1 in a million!'
This means the probability is **nearly** 0%!

What other things have a probability of nearly 0%?

Sometimes you can work out a probability.

Suppose you roll a dice. What is the probability you roll a 1?

There are 6 different numbers you could roll.

If the dice is a fair one the probability of getting a 1 is 1 in 6, or $\frac{1}{6}$ as a fraction.

B3 (a) Use a calculator to work out $\frac{1}{6}$ as a decimal.

(b) What is $\frac{1}{6}$ as a percentage, roughly?

B4 What is the probability that you get a 2 when you roll a dice? (Write your answer as a fraction and then as a percentage.)

B5 What is the probability that you get an even number when you roll a dice? (Write the probability as a fraction and as a percentage.)

B6 In a raffle, there are exactly 100 tickets, and 1 prize. Suppose you buy 1 ticket. Your chance of winning is 1 in 100.

(a) What is this as a fraction? What is it as a percentage?

(b) Suppose you buy 5 tickets. What is the probability you win the prize now?

B7 There are 52 cards in a pack. Suppose you pick 1 card from the pack.

(a) What is the probability that it is a heart? (Write the probability as a fraction, then as a percentage.)

(b) What is the probability that it is an ace?

B8 Which is more likely,
you get a 6 when you roll a dice
or you get an ace when you cut a pack of cards?

c Experimental probability

When you toss a coin it can land *heads* or *tails*.
If the coin is a fair one, you can see that a head
or a tail is equally likely to happen.

So the probability that a coin lands heads is $\frac{1}{2}$ or 50%.

When you throw a drawing pin it can land: *point up*   or   *point down*.
But you don't know which is more likely
unless you find out with an experiment.

Have a go

	Tally
Point up	ⵀⵀ ⵀⵀ ⵀⵀ ⵀⵀ ⵀⵀ
Point down	ⵀⵀ ⵀⵀ ‖

Throw a drawing pin 100 times.
Record how many times it lands point up
and how many times point down.

Do you agree that point up and point
down are not equally likely?

You can work out the experimental probability of getting point up.
Suppose you throw the drawing pin 100 times and get point up 41 times.
The experimental probability of point up is $\frac{41}{100}$ or 41%.

C1 (a) In your experiment, how many times did you get point up?

(b) What was the experimental probability of point up
for your drawing pin?

You need the cuboid you made in the last chapter in question A7.

C2 The faces of the cuboid are labelled A, B and C.
If you roll your cuboid like a dice
it will stop with A, B or C on top.

(a) Which do you think is most likely, A, B or C?

(b) Draw a table like this.

Letter	Tally	Frequency
A		
B		
C		

(c) Roll your cuboid 50 times.
In the table, record the results of each roll.

(d) What is the experimental probability of
getting an A when you roll the cuboid?

(e) What is the experimental probability of getting a B?

7 Money matters

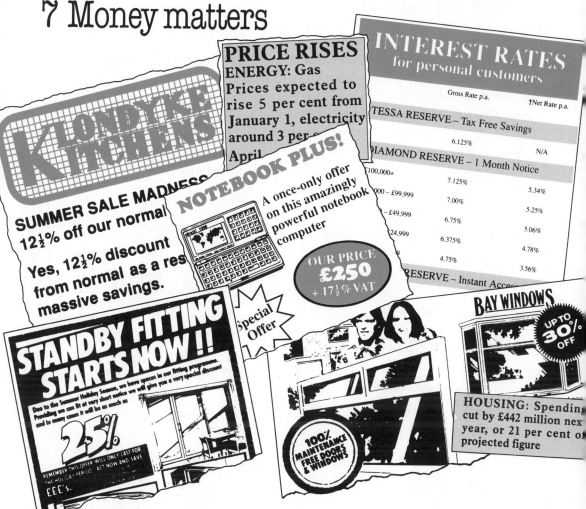

PRICE RISES
ENERGY: Gas Prices expected to rise 5 per cent from January 1, electricity around 3 per c... April

KLONDYKE KITCHENS

SUMMER SALE MADNESS
$12\frac{1}{2}$% off our normal

Yes, $12\frac{1}{2}$% discount from normal as a res... massive savings.

NOTEBOOK PLUS!

A once-only offer on this amazingly powerful notebook computer

OUR PRICE
£250
+ $17\frac{1}{2}$% VAT

INTEREST RATES
for personal customers

	Gross Rate p.a.	†Net Rate p.a.
TESSA RESERVE.– Tax Free Savings		
	6.125%	N/A
DIAMOND RESERVE – 1 Month Notice		
£100,000+	7.125%	5.34%
...000 – £99,999	7.00%	5.25%
... – £49,999	6.75%	5.06%
...24,999	6.375%	4.78%
	4.75%	3.56%
...RESERVE – Instant Acce...		

STANDBY FITTING STARTS NOW !!

Due to the Summer Holiday Season, we have spaces in our fitting programme. Providing we can fit at very short notice we will give you a very special discount and in many cases it will be as much as

25%

REMEMBER THIS OFFER WILL ONLY LAST FOR THE HOLIDAY PERIOD. ACT NOW AND SAVE £££'s.

Special Offer

100% MAINTENANCE FREE DOORS & WINDOWS

BAY WINDOWS

UP TO 30% OFF

HOUSING: Spending cut by £442 million next year, or 21 per cent of projected figure

discussion points

Look at the KLONDYKE KITCHENS advert.
What does discount mean?

Look at the INTEREST RATES leaflet.
What do Gross rate and Net rate mean?

Look at the advert for BAY WINDOWS.
What does 100% maintenance free mean?
There are two adverts for windows.
Can you tell which firm is the cheapest?

The NOTEBOOK PLUS is £249 plus $17\frac{1}{2}$% VAT
How could you work out the VAT?

A VAT and that

VAT is a tax paid to the government.
The rate of VAT changes from time to time.
In 1993 the rate was $17\frac{1}{2}\%$.
You can use a calculator to work out $17\frac{1}{2}\%$.

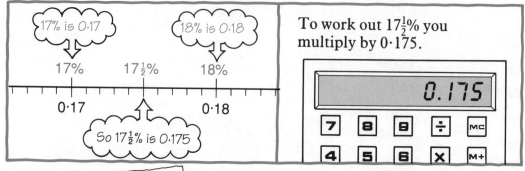

A1 The Notebook Plus computer costs £250 plus $17\frac{1}{2}\%$ VAT.

(a) How much is the VAT?
(Work out $17\frac{1}{2}\%$ of £250.)

(b) How much does the Notebook Plus cost altogether?

A2 Work out the VAT at $17\frac{1}{2}\%$ on each of these.

(a) £180 +VAT.   (b) £82 +VAT   (c) £218 +VAT.

A3 A van costs £6990 + $17\frac{1}{2}\%$ VAT.

(a) How much is the VAT?

(b) How much is the van altogether?

A4 Suppose the VAT rate drops to $12\frac{1}{2}\%$.

(a) How much is the VAT on the van in question A3 now?

(b) How much does the van cost altogether now?

A5 Work out the VAT at $17\frac{1}{2}\%$ on

(a) £26   (b) £498   (c) £42   (d) £199

When you work out $17\frac{1}{2}\%$ of 199
the calculator shows $34\cdot825$.

But £34·825 doesn't make sense.
So you write either £34·82 or £34·83.
You can choose which.

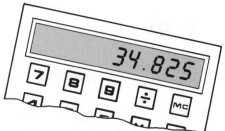

A6 A double glazing salesman gets $22\frac{1}{2}\%$ commission.
For every sale he makes, he gets $22\frac{1}{2}\%$ of the price.
How much commission does he get when he sells

(a) a bay window costing £435

(b) a porch costing £699

(c) a complete set of new windows, costing £2500?

A7 A Dutch bulb company gives 15% discount
when you buy more than 1000 bulbs.
That means they take 15% off the normal price.
Copy and complete each of these bills.

(a)

1000 crocuses	£56·00
15% discount	
Amount to pay	

(b)

2000 daffodils	£120·00
15% discount	
Amount to pay	

(c)

1500 gladioli	£355·00
15% discount	
Amount to pay	

A8 The bulb company gives each worker a bonus in December.
A **bonus** is extra money you would not normally get.
Each worker gets a bonus of 55% of their weekly wage.
How much bonus does each of these workers get?

(a) A packer – weekly wage £126·00

(b) A secretary – weekly wage £180·00

(c) A director – weekly wage £450·00

HEART OF OAK
BUILDING SOCIETY
RATES OF INTEREST
PREMIUM   8% per year
SUPER   11% per year
GOLD $13\frac{1}{2}\%$ per year

A9 The Heart of Oak Building Society
has 3 types of savings accounts.
For example, if you leave £100 in the
Premium account for one year, you get
8% of £100 (£8) interest.
What interest will you get on

(a) £200 put in the Super account for 1 year

(b) £500 in the Gold account for 1 year

(c) £350 in the Premium account for 1 year?
(Hint: 8% is 0·08 as a decimal.)

49

B Borrowing

Bank of Scotland Visa Card

Cardholder's Information Guide

Interest Rates at Date of Issue of this Leaflet

1. Monthly interest rate—2.00%.
2. APR
   (a) In the case of purchases—26.8%

Interest rates and charges

Interest will be charged at 19% per annum (**APR 20.3%**).

Rates will vary from time to time in line with the general level of interest rates and any changes will be announced in branches and in selected newspapers.

There is an annual fee of £10 charged to your Budget Account.

Cheques, Standing Orders and Direct Debits will be charged at the Bank's published personal Current Account tariff. All interest and charges will be debited to your Current Account every three months.

The Budget Account is only available to residents of the British Isles. Midland Bank credit facilities are normally available only to persons of 18 years of age or over.

Apply for a Barclayloan.

BARCLAYS

ANNUAL PERCENTAGE RATE (APR) 19.7%

How to use the Barclayloan Table

A Barclayloan is available for amounts from £300 to £7,500. The table overleaf contains examples of some typical Barclayloans within the range.

The table is very simple to use. For instance on a loan of £1,000 all you have to do is look along the top row of ...

find the re...
the left ha...
most conv...
Where th...
5 figures,...
In this ins...
repaymen...

Access from Midland Bank

Q How much interest do I pay?

A If you settle for your purchases in full each month within 25 days of the statement date then you pay no interest at all

The annual percentage rate of charge is 26.8%. This is based on the maximum annual rate of interest arrived at without making allowance for the time between buying the goods...

BARCLAYS CASHPLAN ACCOUNT.

The following interest rates and entry charges apply from 16th February 1987.

Overdraft interest rates

Overdrawn balances will be charged at a nominal 22% per annum. (This is equivalent to an **annual percentage rate of 23.8%**).

LOCAL LOANS

FOR HOMEOWNERS AND MORTGAGE PAYERS

FAST ANY PURPOSE LOANS £500~£50,000

APR **16.7%** FREE LIFE INSURANCE

36 MONTHLY PAYMENTS	60 MONTHLY PAYMENTS	84 MONTHLY PAYMENT	120 MONTHLY PAYMENTS
20.97	14.46	11.78	–
24.47	16.87	13.74	–
27.97	19.28	15.71	–
31.46	21.70		–
34.96			
43.7			
52.4			
66.4			
87.40			
104.86			

Midland The Listening Bank

PERSONAL LOANS

Examples of Repayments

Flat Rate 10% p.a.

Amount of Loan £	12 MONTHS Annual %Rate **19.5%**			24 MONTHS Annual %Rate **19.7%**		
	Monthly Repayment £	Total Interest £	Total Repayment £	Monthly Repayment £	Total Interest £	Total Repayment £
250	22.92	25.00	275.00	12.50	50.00	300.00
1000	91.67	100.00	1100.00	50.00	200.00	1200.00
3000	275.00	300.00	3300.00	150.00	600.00	3600.00
10000	916.67	1000.00	11000.00	500.00	2000.00	12000.00

discussion points

If you want to borrow money, there can be many ways to do it.
The leaflets above are for loans, credit cards and special bank accounts.
If you borrow money from any of these people
you will have to pay them interest.

Rates of interest can be very confusing!
The **real** rate of interest charged is called APR.
What do you think APR stands for?

Look at the budget account leaflet.
Budget accounts are to help you pay regular bills.
What is the APR?
What else will you have to pay each year?

Which of **these** leaflets has the highest APR?
Why do some people still find this a useful way to borrow money?

What other ways of borrowing money are there?
Find out what the APR is for the other ways.

SAVE
with SPARKLES

CAMCORDERS	Cash ... Cash price	... or easy H.P. terms			APR 29·2%
		Down payment	12 months	(or)	24 months
MITSUBISHI CX1	£449·99	£112·50	£32·22		£18·16
SHARP VLM4H	£499·99	£125·00	£35·80		£20·18
SONY CCDF385	£469·99	£117·50	£33·66		£18·97
JVC GRAX10	£489·99	£122·50	£35·09		£19·78
PANASONIC NVS6	£619·99	£155·00	£44·40		£25·03
CANON E850HI	£649·99	£162·50	£46·55		£26·24

Sparkles sell camcorders.
You can either pay the cash price, or buy on Hire Purchase (H.P.).

If you pay the cash price, you have to pay it all at once.
If you buy on H.P., first you have to pay a down payment.
Then you pay instalments over several months.

B1 Look at the Mitsubishi CX1.

(a) Suppose you buy the CX1
on H.P. over 12 months.
How much is the down payment?

(b) How much is each monthly
instalment?

(c) How much do the 12 monthly
instalments cost altogether?

(d) How much does the CX1 cost
on H.P. altogether?

(e) The Mitsubishi CX1 costs £449·99 for cash.
How much more do you pay if you buy on H.P.?

B2 Look at the Panasonic NVS6.

(a) What is the down payment?

(b) How much do you pay in instalments
if you buy the NVS6 over 24 months?

(c) How much does the NVS6 cost on H.P.?

(d) How much do you save if you pay cash?

c Profit and loss

Tasneem buys a motor-bike for £270. 	She spends £35 on new parts, and sells it for £400. 	Tasneem makes a **profit** of £95 on the bike. 

C1 Tasneem buys another bike for £375.
She spends £56·50 on new parts for it,
and £3·50 on advertising it for sale.
She sells it for £520. How much profit does she make?

C2 Pat repairs clocks.
She buys an old clock for £15, and repairs it.
The parts cost her £1·75.
She sells it for £65·50. How much profit does she make?

C3 Jo buys a car for £675.
He spends £150 on parts.
When he sells it, he only gets £600.

(a) How much did he spend on
buying the car and parts altogether?

(b) Jo makes a **loss** on the car.
His loss is how much he spent altogether,
less what he got when he sold it.
How much was his loss?

C4 A bed shop buys a brass bed for £400.
The shop wants to make a profit of 60% when it sells the bed.

(a) How much profit does it want to make?

(b) For how much must it sell the bed to make this?

(c) The shop sells the bed in a sale for £520.
How much profit does it actually make?

C5 Imran buys some old books for £50.
He hopes to make 50% profit when he sells them.

(a) How much does he hope to sell them for?

(b) He actually sells them for £60.
What percentage profit does he actually make?

D Wages

Most jobs pay people by the hour.

Kitchen Assistant
For small nursery £3·55 per hour; 20 hour week. Friendly atmosphere, possible full time job available later in year.

TRAINEE CHEF
In country restaurant £4·20 per hour for 35 hour week. Interest in food...

PAINTER WANTED
£5·20 an hour for basic 36 hour week. Possible overtime at time and a half. Must be ...enced. Good ...ces essen-...

For example, the kitchen assistant gets paid £3·55 an hour.
If the assistant works 20 hours, he or she will earn

$$20 \times £3·55 = £71·00$$

D1 The trainee chef gets £4·20 an hour.
How much will the chef earn for a 35-hour week?

D2 How much will the painter earn for a 36-hour week?

In some jobs you get paid extra if you work overtime.
Overtime is when you work longer hours than normal.

If the painter works more than 36 hours in a week
the job pays overtime at **time and a half**.

'Time and a half' means you get paid $1\frac{1}{2}$ times the usual rate.
So the painter gets $1\frac{1}{2} \times £5·20 = £7·80$ an hour for overtime.

D3 How much will the painter get paid for 5 hours overtime?

TILER WANTED
Tiler required by old established firm. 35 hour week. Basic rate £4·80 per hour. Overtime at time and a half. Good working conditions. Phone 021 456 7895.

D4 This advert is for a tiler.
'Basic rate' means the normal rate for the job.

(a) How much would you earn
if you worked a 35-hour week?

(b) Overtime is paid at time and a half.
How much an hour is that?

(c) How much would you earn for 6 hours overtime?

(d) Lorraine works for 45 hours one week.
She works 35 hours at basic rate and 10 hours overtime.
How much does she earn that week altogether?

D5 Another firm pays a basic rate of £4·50 for a 35-hour week.
But they pay **double-time** for overtime.
That means they pay double the basic rate for overtime.

(a) How much does this firm pay for 10 hours overtime?

(b) How much would Lorraine earn here for a 45-hour week?

(c) How much would she earn for a 39-hour week?

E Wage packets

When you get paid, you get a pay slip or a wage packet.
The figures on it show what you earned and what
you had to pay in deductions (tax etc.).

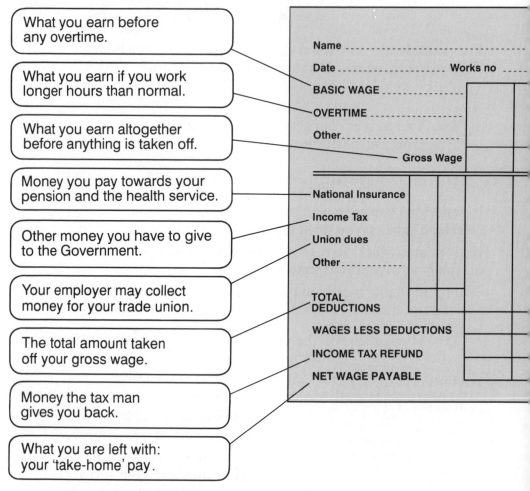

What you earn before any overtime.

What you earn if you work longer hours than normal.

What you earn altogether before anything is taken off.

Money you pay towards your pension and the health service.

Other money you have to give to the Government.

Your employer may collect money for your trade union.

The total amount taken off your gross wage.

Money the tax man gives you back.

What you are left with: your 'take-home' pay.

Name ....

Date .... Works no ....

BASIC WAGE ....

OVERTIME ....

Other...........................

Gross Wage

National Insurance

Income Tax

Union dues

Other...........

TOTAL DEDUCTIONS

WAGES LESS DEDUCTIONS

INCOME TAX REFUND

NET WAGE PAYABLE

E1 Barry is paid £5·60 an hour for a 34-hour week.

(a) What is his basic wage for 34 hours?

(b) He gets paid time and a half for overtime.
How much does he get for 4 hours overtime?

(c) What is his gross wage for a 38-hour week?

E2 When he works 38 hours, Barry pays £13·25 National Insurance,
£37·32 in income tax and £1·40 in union dues.
What are his total deductions?

E3 What is Barry's net wage when he works 38 hours?

Work out the missing figures on these pay packets.

E4

Name: Nick Usiyel
Date: 15/7/94   Works no: 125

BASIC WAGE	261	77
OVERTIME		
Other		
Gross Wage	261	77

National Insurance	15	92
Income Tax	40	41
Union dues		
Other		
TOTAL DEDUCTIONS	(a)	
WAGES LESS DEDUCTIONS		(b)
INCOME TAX REFUND		
NET WAGE PAYABLE		(c)

E5

Name: Gaston Dawson
Date: 15/7/94   Works no: 651

BASIC WAGE		
OVERTIME	185	70
Other	61	20
Gross Wage	(a)	

National Insurance	14	84
Income Tax	36	66
Union dues	2	40
Other		
TOTAL DEDUCTIONS	(b)	
WAGES LESS DEDUCTIONS		(c)
INCOME TAX REFUND		
NET WAGE PAYABLE	(d)	

E6

Name: Izak Glover
Date: 15/7/94   Works no: 379

BASIC WAGE	375	26
OVERTIME	88	41
Other		
Gross Wage	(a)	

National Insurance	27	33
Income Tax	90	66
Union dues		
Other		
TOTAL DEDUCTIONS	(b)	
WAGES LESS DEDUCTIONS		(c)
INCOME TAX REFUND	2	63
NET WAGE PAYABLE	(d)	

E7

Name: Ruth Bastiaens
Date: 15/7/94   Works no: 161

BASIC WAGE	118	40
OVERTIME	6	20
Other		
Gross Wage	(a)	

National Insurance	7	28
Income Tax	14	42
Union dues	1	50
Other	2	00
TOTAL DEDUCTIONS	(b)	
WAGES LESS DEDUCTIONS		(c)
INCOME TAX REFUND	1	07
NET WAGE PAYABLE	(d)	

Review: the garden

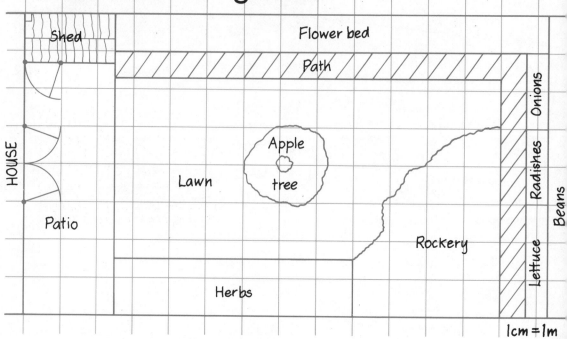

This is the plan of a small garden.

1 The scale of the plan is 1 cm to the metre.
   (a) How long is the whole garden?
   (b) How wide is it?
   (These are called the **dimensions** of the garden.)

2 (a) What are the dimensions of the patio?
   (b) What is the area of the patio?

3 Look at the part of the garden marked Beans
   The owner is going to plant one row of beans.

   CLIMBING FRENCH BEA
   Blue Lake (White Seeded)
   Height approx 150cm (*5ft*)
   Sowing: April–June outdoors. 50mm (*2in*)
   deep in rows approx. 90cm (*3ft*) apart.
   Space seeds 15–23cm (*6–9in*) apart.

   (a) About how many beans will they be able to plant?
   (b) On average, each bean plant will produce about 4 lb of beans. Roughly what will the beans weigh from all the plants?

4 These are the prices of bean poles.
   (a) How many bean poles will the owner need?
   (b) What size is best to use?
   (c) How much will they cost?

Bean Poles (10)	
4 foot	£1·20
5 foot	£1·50
6 foot	£1·80
7 foot	£1·90
8 foot	£2·10

5 Look at the part of the garden
marked *Lettuce*

(a) How many rows could you plant?

(b) About how many lettuces
could you grow here?

LETTUCE

SOWING. April–July, in open ground, 13 mm (½inch) deep, in rows 38 cm (about 15 inches) apart. Thin to 30 cm (about 12 inches) apart.

6 Look at the part marked *Path*

(a) How wide is this part, in centimetres?

(b) This is a price list for paving slabs.
Which size is best to use to
lay a path in this garden?
(1 foot is about 30 cm.)

(c) Draw a rough sketch of the path.
Draw in its dimensions.

(d) Work out how many paving slabs
you would need to make the path.

(e) How much would they cost, **including** the $17\frac{1}{2}$% VAT?

Slabs (1½ in thick)

2ft × 3ft	£2·10 ea.
2ft × 2ft	£1·35 ea.
2ft × 1ft	£1·06 ea.

plus $17\frac{1}{2}$% VAT

7 Instead of laying paving slabs, you could lay
ready-mixed concrete on the path.
Suppose you wanted to lay the concrete about 10 cm deep.

(a) Work out the volume of concrete you would need.
(Think of the concrete as two thin rectangular blocks.)

(b) The concrete costs £73·19 for $1\,m^3$ or £232·18 for $6\,m^3$ (inc. VAT).
How much would it cost to concrete the path?

(c) Which is cheaper, paving slabs or ready-mixed concrete?

8 (a) The lawn covers the largest area of the garden.
Roughly what is the area of the lawn in m^2?
(The tree grows out of the lawn, and there is lawn under it.)

(b) How much would it cost
to cover the lawn with fertiliser?

(c) Suppose you want to cover the lawn
with fertiliser more than once.
Which of these packets gives you
more for your money?

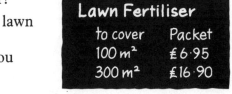

Lawn Fertiliser

to cover	Packet
100 m²	£6·95
300 m²	£16·90

9 The apple tree is very young.
When it is fully grown, the diameter of
its branches will be about 15 feet.
It should be at least 30 feet from the nearest building.

(a) Draw a sketch of the garden and tree.
Show the apple tree when it is fully grown.

(b) Is the tree too near the house?

8 Scales and bearings

A Scales

Maps and plans are used for many different reasons.
Different types of maps and plans use different scales.

The plan of a garden might use
a scale like 1 : 200.
This means that **1cm** on the plan
stands for **200 cm** in the garden.

A map in an atlas might use a scale
like 1 : 20 000 000.
1cm on the map stands for
20 000 000 cm in the real world.

How far is
20 000 000 cm?

100 cm is 1 metre.
So divide by 100 to get metres.

20 000 000 ÷ 100

is   200 000 metres ...

1000 metres is 1 kilometre.
So divide by 1000 to get km.

200 000 ÷ 1000

is   200 kilometres.

So 1 cm on the map
is 200 km in the real world.

A1 What distance does 1 cm stand for in each of these scales?
(Give your answer in metres or km where it makes more sense than cm.)

(a) 1 : 20   (b) 1 : 10 000

(c) 1 : 100   (d) 1 : 2 000 000

A2 A wall chart of the world
has a scale of 1 : 30 000 000.

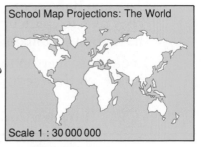

(a) How many km in the real world
does 1 cm on the wall-chart stand for?

(b) From New York to Moscow is 25 cm
on the wall-chart.
Work out how many km it is really.

A3 Here are six different scales and six examples where one could be used.
Pick the best example for each scale.
The first is done for you.

A   1 : 200
B   1 : 50
C   1 : 200 000
D   1 : 50 000
E   1 : 10 000
F   1 : 10 000 000

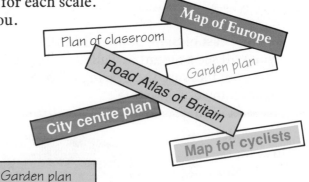

Plan of classroom

Map of Europe

Road Atlas of Britain

Garden plan

City centre plan

Map for cyclists

> A3   A 1 : 200   Garden plan

A4 Measure the distance marked on each of these plans.
Then work out what the real distance must be.

(a) BARKER'S LARGE DOG PEN

Run   Bedding   ?

Scale 1 : 50

(b) Penge Village Hall

?

Scale 1 : 400

59

B Bearings

You need: an angle measurer,
            worksheet G8–4.

Gary and Rakesh go to Wales.
They plan to climb up Snowdon,
the highest mountain in Wales.

Look at the map on the worksheet. The scale of the map is 1 : 50 000.
Find Llyn Nadroedd, a small lake to the west of Snowdon.

> **B1** Llyn Nadroedd is about 3 cm from Snowdon on the map.
> How many km from Snowdon is the real Llyn Nadroedd?

> **B2** There is another lake about 2 km to the east of Snowdon.
> What is its name?

To the east of Snowdon, and a little north
you will see a mountain called Crib-goch.
You can measure the bearing of Crib-goch
from Snowdon like this.

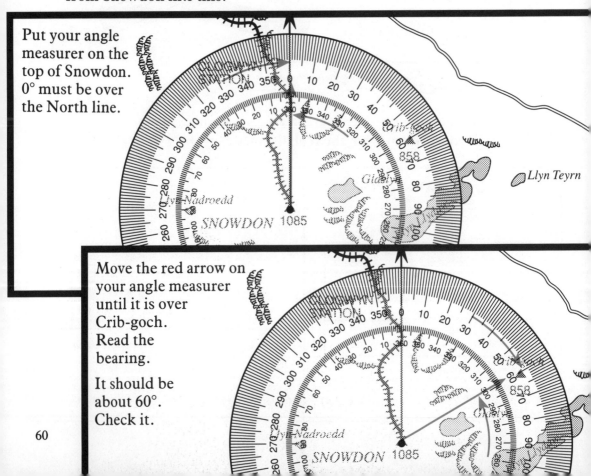

Put your angle
measurer on the
top of Snowdon.
0° must be over
the North line.

Move the red arrow on
your angle measurer
until it is over
Crib-goch.
Read the
bearing.

It should be
about 60°.
Check it.

B3 Measure the bearing of Y Lliwedd from Snowdon.

B4 Copy and complete the bearings column of the table below. You will need to draw some lines on the worksheet to help you find the bearings.

Mountain or place	Bearing from Snowdon	Distance from Snowdon
Crib-goch	60°	
Y Lliwedd		
Gallt y Wenallt		
Pitt's Head		
Foel goch		
Nant Peris		
Clogwyn Station		

B5 On the worksheet, measure the distance of Crib-goch from Snowdon.

Work out how far from Snowdon the real Crib-goch is. Fill in the real distance in the table above.

B6 Measure the distances of the other places from Snowdon. Work out the real distances and fill them in on the table.

B7 The Snowdon Mountain Railway is shown on the map. It runs from Llanberis to the top of Snowdon.

Use the edge of a piece of paper to measure the length of the railway on the worksheet.

How long is the real railway?

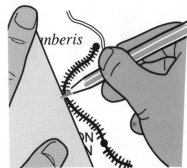

9 Averages

A Mean and median

There are 9 people in the Abacus Keep-fit class.

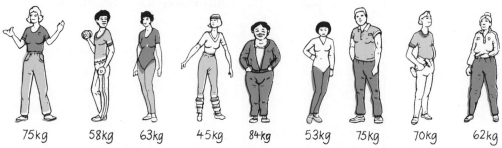

75kg 58kg 63kg 45kg 84kg 53kg 75kg 70kg 62kg

The picture shows the 9 people in the class with their weights.

One way to find the average weight of the class
is to add up all the weights and divide by the number of people.

75 + 58 + 63 + 45 + 84 + 53 + 75 + 70 + 62 = 585. 585 ÷ 9 + <u>65 kg</u>.

> We call this type of average the **mean** weight of the class.

Another way to find the average weight is to put the weights in order.

45kg 53kg 58kg 62 kg 63kg 70kg 75 kg 75 kg 84 kg

Then ask the middle person
what their weight is.

I weigh 63 kg.

> We call this type of average the middle or **median** weight.

62

A1 The ages of the people in the Keep-fit class are:
    19 23 53 19 16 26 77 19 27

    (a) Work out the mean age of the class.

    (b) Write the ages in order, youngest first.
        What is the median age of the class?

In the Keep-fit class, most people are younger than the mean age.
The mean age can be misleading when there are one or two
very old or very young people in a group.

Suppose someone asks you the average age of the Keep-fit class.
The median gives a better idea of the average than the mean.

A2 There are 7 people in an evening typing class.
Their ages are:
    50 42 16 48 15 50 45

    (a) Work out the mean age.

    (b) What is the median age?

    (c) Which gives the better idea of the average age of the class,
        the mean or the median?

A3 Zara wants to find out how long pet tortoises live for.
She finds out how old 5 pet tortoises were when they died.

2 years   **5 years**   **68 years**   **2 years**   **3 years**

    (a) Work out the mean age of the tortoises when they died.

    (b) Find the median age when they died.

    (c) Zara is asked 'About how long do pet tortoises normally live?'
        What should she say?

A4 Steve wants to know how much his stamp collection is worth.
He asks 4 stamp dealers to tell him its value.

      £150   £140   £60   £175

Work out the median value of the stamps.
(Remember: if there isn't one middle value,
the median is half-way between the two middle values.)

B Using the mean

B1 These boxes of map pins say there are 'approx 100' in each box.

Winston counted the number of pins in 5 boxes.
Here are his results.

106, 105, 98, 104, 102

What is the mean number of pins in Winston's boxes?

B2 (a) On average, how long are the pins? (Look on the box.)

(b) About how long would you expect 20 of them to be, end to end?

B3 The mean weight of a pin is 9·8 grams.
Would you be surprised if 50 pins weighed 400 grams?

B4 Veda wants to work out how much her stamp collection is worth.
She picks out 10 stamps and finds out how much each one is worth.

(a) What is the mean value of the 10 stamps?

(b) What is the median value of the stamps?

(c) Veda has 1000 stamps in her collection.
Use the mean value to estimate what they are worth altogether.

B5 45 of the South Wales Sumo Wrestlers Club are going on tour.
They have booked a coach with Jones' Coach Tours.
Mr Jones is worried about the weight!

He asks 5 wrestlers their weights.
They weigh   120 kg   159 kg   130 kg   165 kg   131 kg.

Mr Jones' coach will carry 6000 kg.
What would you advise?

c Mode

Sometimes we use another kind of average – the mode.

> ### The **mode** is the **most common** value.

Here are the ages of the Keep-fit class again, in order.

16   19   19   19   23   26   27   53   77

The most common age is 19, so this is the mode.
You can also say that 19 is the **modal age**.

You don't need to write the ages in order to find the mode,
but it sometimes helps!

C1   Here are the ages of the people in the typing class again.

50   42   16   48   15   50   45

What is the mode of the ages of the people in the typing class?

C2   Look round at the colour of the hair of the people in your class.
What is the modal colour of people's hair in your class?

C3   Pepe wants to start making organic ice-cream.

(a) He asks 15 people what their favourite flavour is.

Vanilla	Raspberry	Vanilla	Coconut
Raspberry	Rum & Raisin	Lemon	Tutti frutti
Coffee	Strawberry	Raspberry	Toffee
Raspberry	Chocolate	Strawberry	

Pepe only wants to make one flavour to start with.
What flavour should he make?

(b) Pepe wants to sell only one size of tub to start with.
He needs to decide what size to sell.
He asks 10 people what size of tub they usually buy.

500 ml	1 litre	2 litres	750 ml
2 litres	2 litres	500 ml	
1 litre	4 litres	4 litres	

What size of tub do you suggest he makes?

D Range

Pepe starts selling his ice-cream in June.
You can use a number line to show how much money he takes each day.
Each dot represents one day's sales.

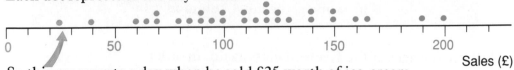

So this represents a day when he sold £25 worth of ice-cream.

The sales range from £25 to £200 a day.

In mathematics, the word **range** has a special meaning.
On the best day, Pepe sells £200 worth of ice-cream.
On the worst day he sells £25 worth.
We say the range is £200 − £25, or £175.

> To find the range, **take the smallest value from the biggest value.**

D1 This number line shows Pepe's sales from July 1st to July 14th.

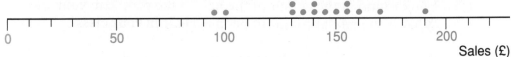

What was the range of his sales from July 1st to the 14th?

D2 Mike sells ice-cream every day in Victoria Park.
This number line shows his daily sales in July.

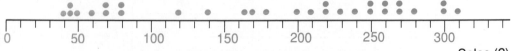

This number line shows his daily sales in August.

(a) What was the range of sales in July?

(b) What was the range of sales in August?

(c) One of these two months was quite warm every day.
The other had some very hot days and some very wet days.
Which do you think was which? Why?

D3 Here are Mike's daily sales in the first two weeks in September.
£135, £90, £220, £295, £120, £330, £135, £120, £180, £125, £265, £200, £220, £295

(a) Write the sales in order, smallest first.

(b) What was the range of his daily sales for these two weeks?

66

D4 Upton Youth Club have to pick a girl to throw the discus.
There are only two girls to pick from.
Here are their throws (in metres) in their last 5 competitions.

Rita   27·6   31·4   32·5   27·5   28·0
Alessa   29·1   28·9   30·1   30·0   28·9

(a) Rita's mean throw for these 5 competitions is 29·4 m.
Check that Alessa has the same mean throw.

(b) What is the range of Rita's five throws?

(c) What is the range of Alessa's five throws?

(d) The club want the more reliable discus thrower.
Who should they pick?
Give a mathematical reason why they should pick this girl.

D5 Durmus drives to work each morning.
He can drive through Todsdon or through Litham.
He checks how many minutes it takes him over each of the two
routes for a few days.

Going by Todsdon   14   18   19   32   17
Going by Litham   17   19   20   21   18   19

(a) On average, which route is quicker? (Use the mean.)

(b) Which route has the bigger range of journey times?

(c) Durmus gets up late. He jumps into his car at 8:46.
Unless he gets to work at 9:00 he will get the sack.
Which route should he choose? Explain why.

D6 Tom and Bert both grow tomatoes.
Tom thinks he grows the heaviest tomatoes. Bert thinks *he* does!

They each weigh 10 of their own tomatoes,
and work out the mean and range of the weights.

(a) Who do you think has the best chance of winning
the Heaviest Single Tomato competition?

(b) If you bought 10 tomatoes off Tom, how much
would you expect them to weigh?

(c) How much would you expect 10 of Bert's tomatoes to weigh?

Review: reflection symmetry

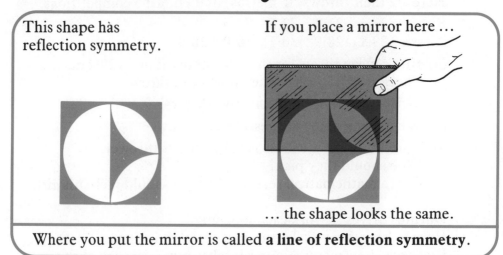

This shape hàs reflection symmetry.

If you place a mirror here ...

... the shape looks the same.

Where you put the mirror is called **a line of reflection symmetry**.

1 Which of these have reflection symmetry?

A

B

C

D

E

F

G

H

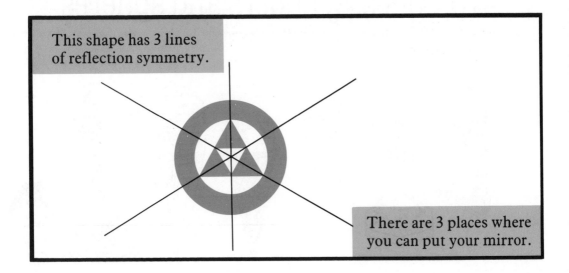

This shape has 3 lines of reflection symmetry.

There are 3 places where you can put your mirror.

2 How many lines of reflection symmetry does each of these shapes have?

(a) (b) (c)

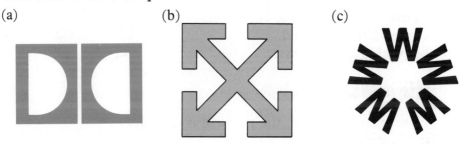

3 Copy each of these shapes onto triangular spotty paper. Add extra shading to make each of them symmetrical about the lines of symmetry shown.

(a) (b) (c)

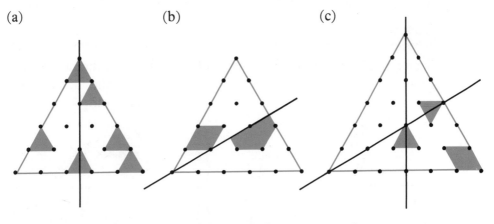

4 On triangular spotty paper, invent some patterns of your own that have more than one line of symmetry.

10 Cones, cylinders and spheres

A Cones

Look at the pictures above.
The everyday word for
all of these things is **cone**.

In mathematics, a cone is
a thing like a pyramid,
but with a circular base.

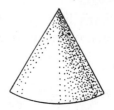

discussion point

None of the things in the pictures is an exact
mathematical cone.
What is wrong with each of them?

Can you think of other things which look like
a cone? (We call things like cones **conical**.)

Making a cone

You need compasses, scissors, paper and glue.

1 Set your compasses to 5 cm.
Draw a circle, radius 5 cm.

2 Draw a diameter on your circle.
Mark one half of the diameter
every $\frac{1}{2}$ cm.

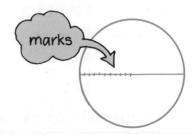

marks

3 Cut along the other half of the diameter.
Only cut to the centre.

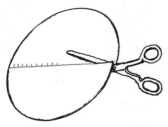

4 Overlap and glue the cut parts to make a cone.
The pencil marks should be inside.

Glue

discussion points

All these cones were made from the same size circles.

What is different about the way each is made?

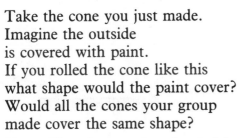

Take the cone you just made.
Imagine the outside is covered with paint.
If you rolled the cone like this what shape would the paint cover?
Would all the cones your group made cover the same shape?

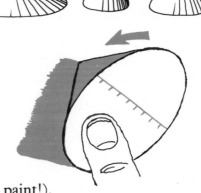

Try rolling them and seeing (without paint!).

This is a picture of 3 cocktail glasses.
One of them is half full.
Which do you think it is?

Try it

You can find the answer to the last question yourself (without the cocktails!).
Fill your cone with sand, or something similar.
Weigh the sand. Then weigh out half the sand.
Pour half the sand back into the cone.
How far up the side does it come on your pencil marks?

B Cylinders

Many things are
cylinder shaped
so that they can roll.

Before people used
wheels, they used
tree trunks shaped like cylinders
to move heavy things about.

discussion points

As the people pull, the rollers move more slowly than the block.
What would the people pulling the block
have to do as they went along?

If you are not sure
try putting some
round pencils under
a book and pushing
it along.

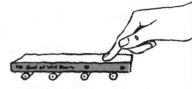

This is a photo of an
old ruler.
It is shaped like a
cylinder.
It was used to draw
parallel lines.
How do you think
it was used?

Most tin cans are shaped
like cylinders.
Look at some cans.
How are they made?
Can you tell what size
pieces of metal your
cans were made from?

An experiment

1 Take an A4 sheet of paper.
Fold it carefully in half.
Then cut it exactly
along the fold line.

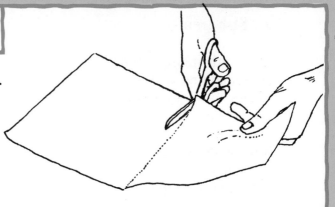

2 Fold one half sheet in four.
Re-fold it to make a triangle
shaped pillar.
Glue one edge of the pillar.

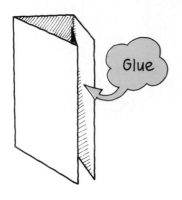

3 Roll the other half
into a cylinder.
Glue down one
edge.

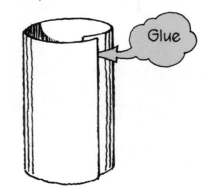

4 Now put G books on
top of each of your
pillars.
Put the books on
very carefully.
Make a note of
how many books each
pillar will hold.

Which pillar is strongest?

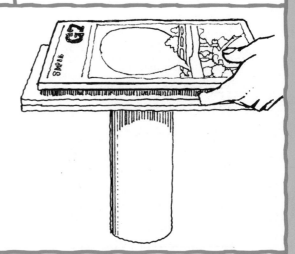

Try making a square pillar out of half an A4 sheet.
How strong is a square pillar?

C Spheres

The mathematical name
for something that is
round, like a balloon,
is a **sphere**.

We use the word when we
talk about the
atmo**sphere** round
the earth.

Some buildings are almost spheres.

—Investigation!—

You need a polygon stencil, a football, a tennis ball,
a cricket ball. . .

Look at the football.
It is made from small flat
pieces of leather.

The pieces are sewn together
at the edges.

Use the polygon stencil to draw
a pattern for the football.
The pieces you draw should join at the edges.
(But there will be some gaps!)

Look at the tennis ball.
It is made from two pieces.

Draw a rough sketch of the two pieces.

Draw sketches of the pieces that the cricket ball
and others are made from.

11 Rates

A Heart rate

When you run, your heart beats faster.
How fast your heart is beating is called your **pulse rate.**

This graph shows the pulse rate of an athlete running a race.
It shows her pulse rate before and after the race as well.

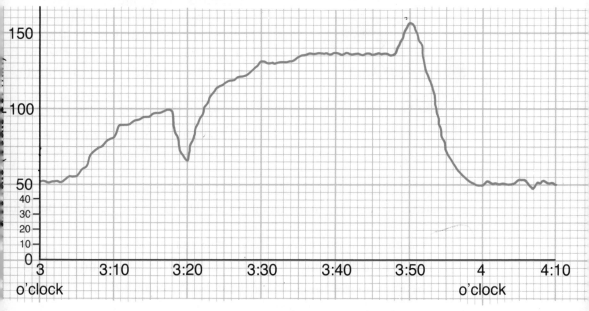

A1 The athlete jogs around before a race, then she
rests for a minute or two before the start.
What time do you think the start of the race was?

A2 (a) What was the athlete's pulse rate
before she started to warm up?

(b) What was her pulse rate at the start of the race?

A3 (a) What time do you think she finished the race?

(b) What was her pulse rate then?

(c) How long did it take the athlete's pulse rate
to go back down to normal?

A4 (a) How long did it take the athlete to run the race?

(b) This runner usually averages about 10 m.p.h.
in races this long.
About how far do you think the race was?

Your pulse rate is measured in **beats per minute.**
The athlete's normal pulse rate was 50 beats per minute.
– every minute her heart beats 50 times.

Different people have different pulse rates.

An average woman (who is **not** an athlete) has
a normal pulse rate of about 80 beats per minute.
This is her pulse rate when she is resting.

A5 This nurse is finding out a
patient's pulse rate.
She counts the number of beats
the boy's heart makes in 30 sec.
Then she doubles the number to fin
his pulse rate in beats per minute.

(a) What is his pulse rate if the
nurse counts 50 beats in 30 sec?

(b) What is his pulse rate if she
counts 44 beats in 30 sec?

An average man has a lower pulse rate than a woman.
When resting, an average man has a pulse rate of about 72 beats per min.

A6 A man has a pulse rate of 72 beats per minute.

(a) How many times will his heart beat in 30 seconds?

(b) How many times will it beat in 15 seconds?

A7 This doctor is timing the girl's pulse
He counts the girl's heart beats
for 15 seconds.

(a) What is the girl's pulse rate if th
doctor counts 22 beats in 15 sec?

(b) What is her pulse rate if he
counts 16 beats?

Test yourself !

Time your heart beat for 15 seconds.
Work out your pulse rate. (Make sure you are resting!)

Now do some exercise – run round for 5 minutes
or step up and down on a chair for a minute or two.

Stop and time your heart beat straight away for 15 seconds.

Then time your heart again 1 minute after you stopped.
Time again after 2 minutes, 3 minutes, and so on.

76 Work out your pulse rates. Draw a graph of them.

B Birth rates

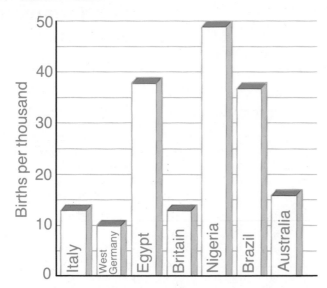

This bar chart shows the **birth rate** in 7 countries in 1983.
The birth rate is shown as **births per thousand**.
Italy had a birth rate of 13 births per thousand.
This means that in Italy in 1983,
13 children were born for every thousand Italians.

B1   (a) Which of these countries had the highest birth rate in 1983?

      (b) Which country had the lowest birth rate?

B2   Pora is an Italian village.
About 3 thousand people live there.
Roughly how many children do you think
were born in Pora in 1983?

B3   About half a million people live in Leeds (Yorkshire).

      (a) How many thousands is half a million?

      (b) Roughly how many children do you think
         would be born in Leeds in a year?

B4   Kano is a large city in Nigeria.
About 400 000 people live in Kano.
Roughly how many children do you think
will be born in Kano each year?

B5   Monaco is a very small country.
Only 26 000 people live there.
About 200 children are born in Monaco each year.
What is the birth rate in Monaco?

C Working out rates

Sometimes you need to work out a rate

This box shows how fast you have to type to pass a typing exam.

For example, to pass Stage 1, you have to type at 25 words per minute.

RSA TYPEWRITING
Examination speeds

Stage 1   25 w.p.m.

Stage 2   35 w.p.m.

Stage 3   50 w.p.m.

To work out a rate like **words per minute** you divide

number of words ÷ number of minutes

C1 Use a calculator to work out the boy's typing rate.
Key in [2] [0] [0] [÷] [6] [=]
(a) **Roughly** what is his typing rate in words per minute?
(b) Can he type fast enough to pass Stage 1?

C2 A girl types 290 words in 7 minutes.
(a) Roughly what is her typing rate?
(b) Does she type faster or slower than the boy in question C1?
(c) Which RSA Stage do you think she could pass?

C3 Which RSA Stage do you think each of these people could pass?
(a) Ranjit typed a letter in 8 minutes. The letter had 300 words in it.
(b) Dawn typed a test for her RSA exam. The test had 216 words in it, and Dawn typed it in exactly 4 minutes.

C4 Work out your own **writing** rate.
Pick a page of a book, then see how many words you can write in exactly 5 minutes.

Some cars use more petrol than others.
To compare different cars, you can work out
how many **miles per gallon** a car uses.

You work out $\dfrac{number\ of\ miles\ car\ goes}{number\ of\ gallons\ it\ uses}$

(Remember the line across means you divide.)

C5   A car travels 200 miles and uses 4 gallons of petrol.
How many miles per gallon did it do?

C6   On holiday, the Smiths
did 1206 miles.
They used 27 gallons of petrol.
About how many miles per gallon
did their car do?

C7   Most petrol pumps show how much
you buy in **litres.**
To convert from litres to
gallons you can use this
ready reckoner.

(a) How many gallons is 20 l?

(b) How many gallons is 25 l?

(c) How many gallons is 49 l?

Litres	Gallons	Litres	Gallons
1	0·22	10	2·20
2	0·44	20	4·40
3	0·66	30	6·60
4	0·88	40	8·80
5	1·10	50	11·00
6	1·32	60	13·20
7	1·54	70	15·40
8	1·76	80	17·60
9	1·98	90	19·80

C8   Seth takes his car to London and back.
He travels 126 miles.
This pump shows how much petrol he uses.

(a) How many **litres** of petrol did he use?

(b) How many gallons is that?

(c) About how many miles per gallon did his car do?
Remember: divide miles by gallons

Litres	18
Price £	9.72
Price per litre £	0.54

C9   The pictures show my speedo before and after a trip.

(a) How many miles did I travel?

(b) I used 21 litres of petrol.
How many miles per gallon did my car do?

79

Review: congruence

You need tracing paper.

These two keys are **congruent**.

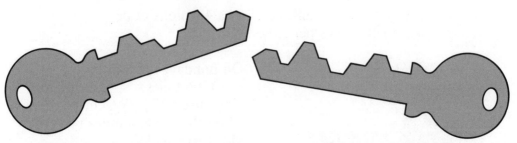

If you cut them out, you could fit one exactly on top of the other.
They are both exactly the same shape and same size.

1   Look at this key.
    It is congruent to some of the
    keys below.

    Which keys is it congruent to?
    (Tracing paper might help!)

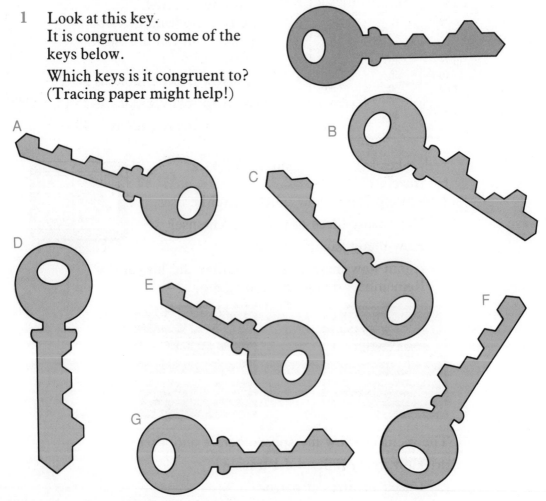

2 Look at the shapes on this page.
There are some pairs of congruent shapes.
Which are they?

For example, if you think shape A is congruent to shape B
write *A is congruent to B.*

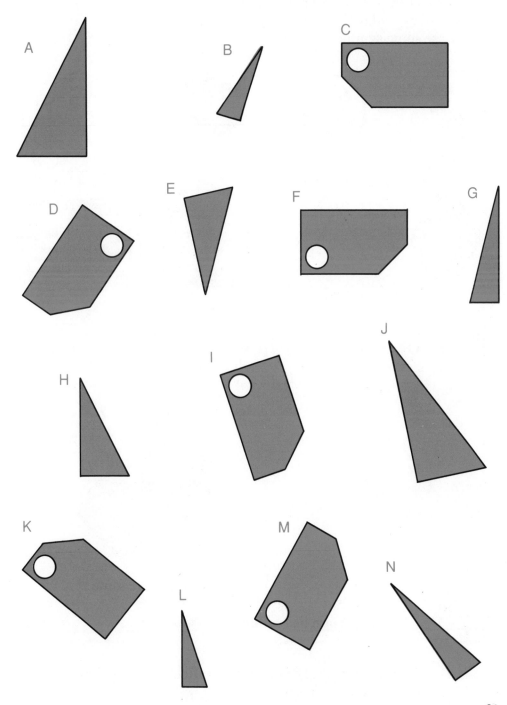

12 Rotation Symmetry

A Everyday symmetry

People like things to be symmetrical.
So lots of everyday things have symmetry.
Here are some symmetrical designs.

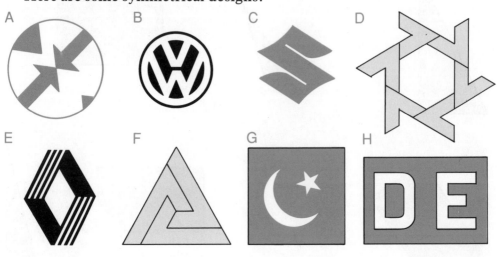

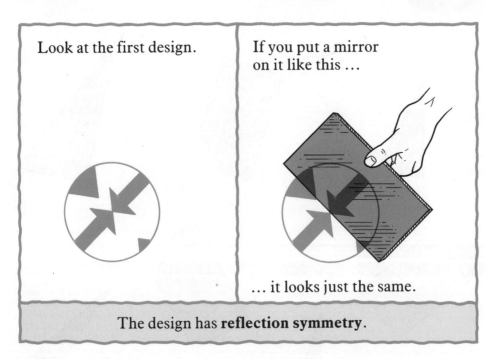

Look at the first design.

If you put a mirror on it like this ...

... it looks just the same.

The design has **reflection symmetry**.

A1 Some of the other designs have reflection symmetry.
List the ones that do.

82

Here is design C – a Suzuki badge. The Suzuki badge does not have reflection symmetry.

Wherever you put a mirror …

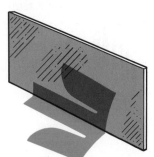

… it looks different.

But if you turn the page upside down the Suzuki badge looks exactly the same.

> We say the Suzuki badge has **rotation symmetry**.

It looks the same in 2 positions, normal way up and upside down.

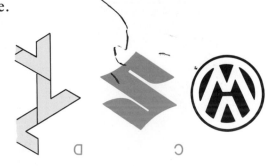

Ⅾ   Ɔ

Look at design F. It doesn't have reflection symmetry.

But it does look the same if you turn the page. It does have rotation symmetry.

A2 In how many different positions does design F look the same? (Remember to include the original position.)

> A shape with rotation symmetry looks the same in different positions. The number of positions it looks the same in is **the order of rotation symmetry**.

A3 (a) What is the order of rotation symmetry of the Suzuki badge?
(b) What is the order of rotation symmetry of design F?

A4 (a) Which other shapes on page 82 have rotation symmetry?
(b) What is the order of rotation symmetry of each of them?

A5 Some of these shapes have rotation symmetry. Some do not.

For each shape, say whether it has rotation symmetry or not.
For the ones that do, say what their order of rotation symmetry is.

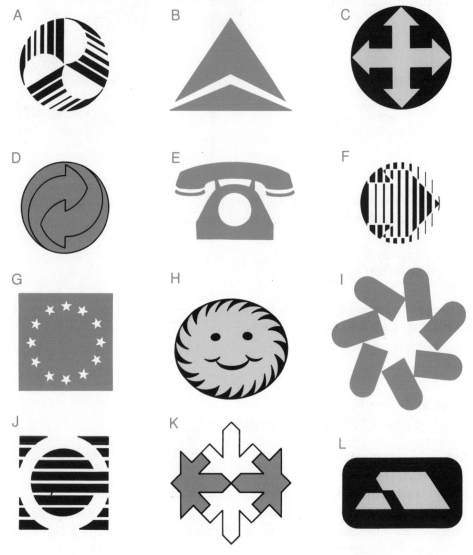

A6 Some letters have rotation symmetry.
Which of these letters have rotation symmetry?
For the ones that do, say what their order of rotation symmetry is.

A B H I N S X

A7 Some of the letters in question **A6** have reflection symmetry.
Which of the letters have reflection symmetry?

B Centres of rotation

You need worksheet G8–5, tracing paper and a sharp pencil.

It can be useful to use tracing paper to find rotation symmetry.

1 Look at shape A on worksheet G8–5.

2 Make a tracing of the shape. Put a cross at its centre.

3 Put your pencil at the centre. Turn the tracing paper until it fits over the original.

4 Turn the tracing paper again until it fits over the original.

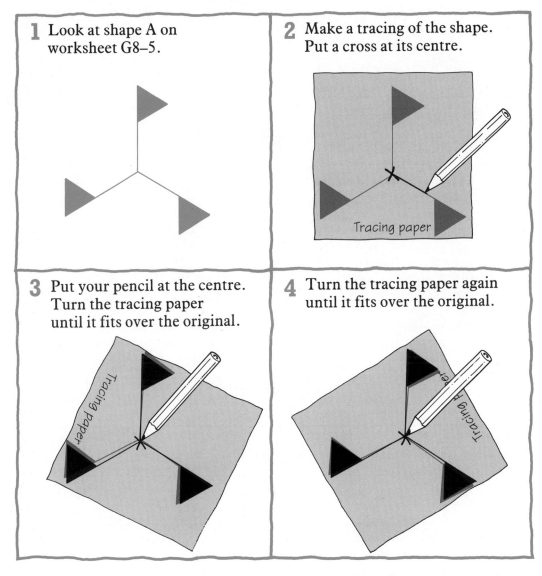

The tracing paper can fit in three positions over the original (including the starting position).

So shape A has rotation symmetry of order 3.

> The place you put your pencil to rotate the tracing paper is called **the centre of rotation**.

B1 Look at shapes B to E on the worksheet.
Use tracing paper to find the order of rotation symmetry
of each shape.

Mark the centre of rotation on each shape.

Look at this shape.
At the moment it does not
have rotation symmetry.

But you can shade more
of the shape to make
it have rotation symmetry.

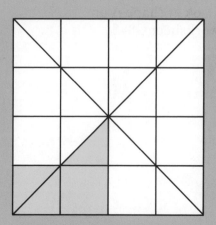

You can shade so it has
rotation symmetry of order 2.

Or you can shade so it has
rotation symmetry of order 4.

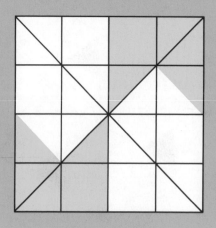

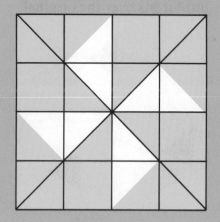

B2 Shade in the shapes on the worksheet.
Make each of them have the order of rotation symmetry shown.

B3 There are some partly-finished designs on the worksheet.
Finish them so they have rotation symmetry.

B4 Draw some shapes of your own with rotation symmetry.
For each one, say what its order of rotation symmetry is.

Review: chapters 1 and 2

1 What is the volume of each of these boxes?

(a)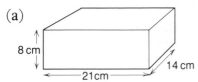

(b)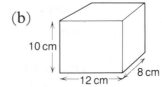

2 Each of these boxes holds 1 litre.

(a) Work out the volume of each box.

(b) Which is nearest to 1 litre?

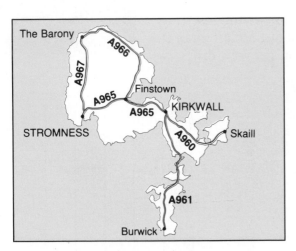

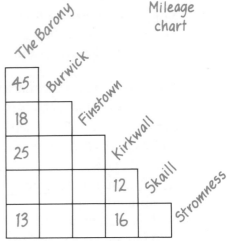

3 The map shows the main roads on the Orkney Islands.

(a) Draw a rough sketch of the roads shown on the map.

(b) Copy the mileage chart.

(c) How far is it from the Barony to Finstown?
Copy the distance on your map.

(d) How far is it from the Barony to Kirkwall?

(e) How far must it be from Finstown to Kirkwall? Put it on your map.

(f) Work out the other distances. Finish the map and mileage chart.

Review: chapters 3 and 4

1   Write these in figures.
    (a) Two hundred thousand
    (b) One and a half million
    (c) Half a million pounds
    (d) A hundred and five thousand

2   Write these in order, largest first.

    £365,219   £2½m   £193,216·53   £200 thousand

3   Write a shorter headline for each of these.

    (a)
    £2,165,396 WON!

    (b)
    £1,509,312 FRAUD

    (c)
    £10,596,312 SHARED!

    (d)
    4,976,132 JOBLESS

4   Maxiflops claim you can store 1000 pages
    on one of their disks.

    Suppose a page has 60 lines of writing on it.
    On average, each line has about 50 letters in it.
    How many letters would you get on a disk?

    MAXI FLOPS
    Computer floppy disks
    STORE 1000 PAGES
    ON 1 DISK!!

5   This chart shows the marks
    some children got in a test.

    For example, you can see
    that 2 children got less
    than 50 marks.

    (a) How many children
        altogether got
        less than 60 marks?

    (b) How many got at
        least 70 marks, but
        less than 90 marks?

    (c) How many children
        took the test altogether?

    (d) To pass the test, you
        need at least 60 marks.
        How many children passed the test?

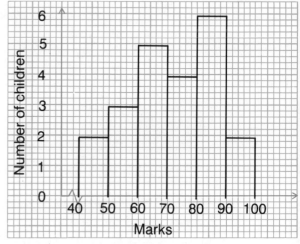

Review: chapters 5 and 6

1 Which of these are nets of cuboids?

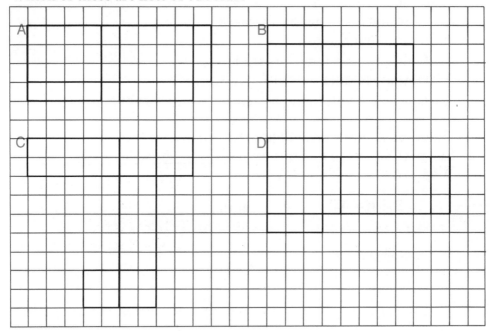

2 Rewrite these sentences in ordinary English!

I think that the probability that your team will win the cup is 0%.

I think that the probability that my team will win the cup is 100%.

3 (a) There are 52 cards in a pack.
   If you pick a card, what is the probability that it is a club?

   (b) What is the probability it is a king?

4 This board is a kind of raffle.
   You pick a square, and rub off the top.
   If it says PRIZE underneath, you win!
   25 of the squares say PRIZE underneath.

   PICK A LUCKY SQUARE

   (a) What is the probability of winning
   if you pick one square?

   (b) 30 people pick squares.
   About how many do you expect to win?

Review: chapters 7 and 8

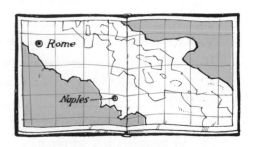

1   Laura gets a job as a plumber.

   (a) How much will her basic wage be
       for a 38-hour week?

   (b) Laura works 5 hours overtime.
       How much does she get for this?

   (c) How much will Laura earn if she works 48 hours in a week?

2   One week Laura works 40 hours.
    She pays £23·32 in income tax and £11·24 National Insurance.
    What is her take-home pay?

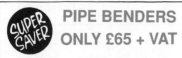

PIPE BENDERS
ONLY £65 + VAT

12½ % trade discount

3   Laura buys herself a pipe bender.
    The normal price is £65.
    Laura gets a 12½% discount.

   (a) How much discount does she get?

   (b) How much does the pipe bender cost her before VAT?

   (c) She has to pay 17½% VAT. What price does she pay altogether?

4   The scale of this plan is 1 : 200.

   (a) What distance does 1 cm
       on the plan stand for?

   (b) How wide are the doors?

   (c) How long is the garage?

   (d) How wide is it?

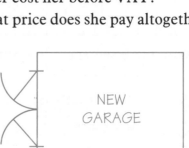

NEW
GARAGE

   (e) It will cost £8·50 per square metre to concrete the floor.
       How much will that cost altogether?

5   What distance does 1 cm stand for in each of these scales?

   (a) 1 : 25 000   (b) 1 : 500   (c) 1 : 500 000

6   In an atlas, a map has
    a scale of 1 : 1 000 000.

    On the map, the distance between
    Rome and Naples is 19 cm.

    How many kilometres is it
    really between Rome and Naples?

Review: chapters 9 and 10

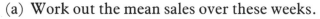

1 The Lucky Bingo Club has a jackpot each week.
   Over 7 weeks, the jackpot was
   £30, £20, £40, £50, £40, £9000, £60

   (a) What was the mean jackpot?

   (b) What was the median jackpot?

   (c) What was the range of jackpots?

2 Giorgio owns an ice-cream van.
   His sales each week in August are
   £820, £690, £920, £870

   (a) Work out the mean sales over these weeks.

   (b) If his sales continued at the same rate,
       how much would his **total** sales be
       after 52 weeks?

   (c) Why would this not be a sensible way
       to work out what he would expect to sell in 1 year?

3 Write down the mathematical name
   of each of these shapes.

   (a)

   (b)

   (c)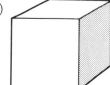

4 (a) Write down the names of 2 ordinary objects
       that are shaped like a sphere.

   (b) Write down 2 things that are shaped like cylinders.

5 Suppose you put a sphere
   into a rectangular box
   that just fits it.
   The sphere takes up
   about 52% of the
   volume of the box.

   Roughly what is the volume
   of this sphere?

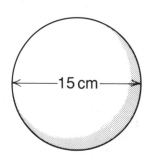

15 cm

Review: chapters 11 and 12

1   In air, sound travels at about 330 metres per second (m/s).

     (a) About how many seconds will sound take
         to travel 1 km?

     (b) How far would sound travel in 10 seconds?

2   In a storm you see lightning straight away.
     The sound of the thunder takes time to
     get to you, because it travels more slowly.

     (a) Suppose you hear thunder 5 seconds
         after seeing the lightning.
         How far away is the storm?

     (b) How far away is the storm if it takes
         20 seconds for you to hear the thunder?

3   A room measures 20 feet by 16 feet.
     Winston wants to lay tiles on the floor.
     The tiles are exactly 1 foot by 1 foot,
     and he can lay them at a rate of 20 an hour.
     About how long will it take him to lay the tiles?

4   Some of these shapes have rotation symmetry. Some do not.
     Say which do have rotation symmetry and
     what their order of rotation symmetry is.

A

B

C

D

E

F